Giacomo Torzo

OPERATIONAL AMPLIFIERS:
BASIC CONCEPTS and COOKBOOK

An experimental approach to analog electronics,

with a short introduction to digital IC and sensors,

for practical applications

Edition 2012

Printed by "Lulu.com"

Foreword

Understanding Integrated-Circuit (IC) electronics is a "brain-tool" that is becoming important in a growing number of scientific studies. However the student frequently feels the first approach to this discipline as a shock. Several textbooks in fact require that the reader invest a great effort before the benefit/cost ratio becomes favorable.

For example often the textbook starts with a difficult and discouraging introduction on transistors. The transistor is indeed the basic element in any IC, but learning its working principle is not necessary for learning IC.

In the modern *analog* electronic circuits, on the other hand, the basic building block is now the Operational Amplifier (OA), not the transistor. And understanding the AO is much easier than understanding the transistor.

Therefore here we start describing the AO and its most important applications, leaving a simplified description of diodes and transistors behavior in an optional Appendix (because in some special circuit the transistor must be used by itself).

The goal of this book is to help the first steps of the students (mainly those whose main interest is *not* electronics) to acquire familiarity with the essential elements of analog electronics, making possible the understanding of many practical circuits.

Algebra is the only mathematical tool strictly required: an elementary knowledge of derivative and integral is enough. Reading the short resume of the complex number properties and of the Laplace transform, in Appendix, should make faster the analysis of the circuits treated in the chapter devoted to filters.

This first English edition of the book is mostly a translation from the original Italian version (published by Decibel-Zanichelli Eds., 1991), with some updates.

This book collects ideas selected from many sources and suggestions of many authors, so that the complete list of people to which I am indebted would be extremely long; but I cannot omit to acknowledge the main help received by Lorenzo Bruschi, and the useful proof-reading made by Giorgio Delfitto.

GIACOMO TORZO

Padova, august 2012

Index

How to use this book 1

1. Introduction
1.1. Voltage and current signals 2
1.2. Resistors, capacitors, inductances, signal sources 2
1.3. Linearity, superposition, Kirchhoff's laws 3

2. Operational amplifiers
2.1. Basic concepts and definitions 6
2.2. Ideal OA 9
2.3. Real OA 9

The operational amplifier as signal processor
3.1. Inverting amplifier 11
3.2. Non-inverting amplifier 12
3.3. Voltage follower 12
3.4. Differential amplifier 13
3.5. Inverting summer 14
3.6. Non-inverting summer 14
3.7. Effects of bias currents and offset 15
3.8. Effect of the finite open loop gain 15
3.9. Input impedance and output impedance in a closed loop 16

4. Some examples
4.1. Differential with variable gain 20
4.2. Differential with linear gain control 21
4.3. Differential with gain control and high Z_{in} 23
4.4. Instrumentation amplifier 24
4.5. Amplifier with linear gain control from –K to +K 25

5. Reference voltage source
5.1. Zener in the feedback 26
5.2. Dual voltage source 27

6. Voltage to current converter
6.1. Floating load 28
6.2. Floating supply 29
6.3. Floating control signal 29
6.4. Voltage-controlled current source with all signals referred to ground 30
6.5. Full reference to ground using 2 AO 31
6.6. Current source with potentiometric control 32

7. Non linear circuits
7.1. Half-wave rectifier 34
7.2. Full-wave rectifier 35
7.3. Peak detector 38
7.4. Logarithmic and exponential amplifiers 39

8. Active filters
8.1. Integrator 42
8.2. Differentiator 44
8.3. Multiple feedback filters 45
8.4. Quality factor and damping factor 47
8.5. VCVS filters 49
8.6. State variable filters 51
8.7. Simple notch filter 53
8.8. Impedance converter 53
8.9. Gyrator 54
8.10. Capacitance multiplier 55
8.11. IC active filters 56

9. Switching circuits
9.1. Comparator 58
9.2. Comparator with hysteresis 58
9.3. Bipolar astable multivibrator 60
9.4. Unipolar astable multivibrator 61

10. Self-oscillation
10.1. General remarks 63
10.2. Wien-bridge sinusoidal oscillator 64
10.3. Phase shifter 65
10.4. Double shifter oscillator 66
10.5. Quadrature shifter 67
10.6. Double integrator oscillator 67
10.7. Phase shift oscillator 68
10.8. Square/triangular wave generator 69
10.9. Quadrature square/triangular wave generator 71
10.10. Voltage to frequency converter 71
10.11. Frequency to voltage converter 72

11. Phase-sensitive detector (lock-in)
11.1. Lock-in as synchronous switch 74
11.2. Lock-in with multiplier 75
11.3. Lock-in with multiplier ±1 77
11.4 Synchronous filter 79

12. Digital electronics: elementary notions
12.1. Logic circuits 81
12.2. Bistable circuits: the flip-flop 86
12.3. Synchronous flip-flop 87
12.4. Monostables 89
12.5. Astables 90
12.6. Monostable with delay 92
12.7 Delay generators 93

13. Some special IC
13.1. The timer 94
13.2. Integrated voltage sources 98

13.3. Analog switches 100

14. Transducers and interfacing techniques
14.1. Temperature sensors 102
14.2. Force and pressure sensors 109
14.3. Light sensors 112
14.4. Position sensors 117

15. The OA with double feedback 120

16. Guide to experiments
16.1. Some preliminary suggestions 122
16.2. Exercises 123

Appendix A
A.1. The diode 138
A.2. The Zener diode 140
A.3. The transistor : some definitions 141
A.4. Common emitter configuration 142
A.5. Dynamic regime 143
A.6. Common collector (emitter-follower) 145
A.7. Field Effect Transistors (FET) 146

Appendix B
B.1. Complex numbers 148
B.2. Sinusoidal voltages and currents in complex notation 148
B.3. Complex impedance 149
B.4. Complex transfer function 150
B.5. Bode diagram 150
B.6. Laplace transform 151

Appendix C
C.1. Resistors 155
C.2. Potentiometers 157
C.3. Capacitors 157
C.4. Inductors 160
C.5. Diodes 161
C.6. Solderless breadboard 162

Appendix D
D.1. Shortlist of linear IC manufacturers 164
D.2. Pin out and general datasheets of OA 165
D.3. Comparators 170
D.4. Basic list of logic gates (TTL and CMOS) 178

Bibliography 172

How to use this book

This book might be used as theoretical guide to understand various applications of integrated circuits, but it was written as *practical guide*.

The first chapter is a mere collection of definitions and rules that will be frequently used

Chapters 2,3 offer a short introduction to the basic OA circuits, and the reader should try experimenting some simple exercises suggested in chapter 16 before proceeding to next chapters.

Next chapters (4-13) give examples grouped by functions: amplifiers (4), voltage sources (5) current sources (6), non-linear circuits (7), filters (8), comparators and pulsers (9), oscillators (10), lock-in (11), digital circuits (12) and timers, IC voltage regulators and analog switches (13). For all these circuits some suggestions for experimental tests are given in chapters 16.

At this point the reader may feel confident to try setting up interfacing circuits for transducers or sensors, thus exiting from the pure "electronic-world" and entering the wider world of "physics-laboratory": chapter 14 offer several examples of simple interfacing circuitry for some physical quantities: temperature, pressure, position, light.

Chapter 15 is devoted to discuss a topic (OA with positive and negative feedback), which is rarely treated in most handbooks, without involving too complex math notations.

Chapter 16 suggests some practical exercises with the circuits described in the previous chapters, giving in most cases only suitable values for the passive elements and sometimes also some hints for performing elementary measurements. The choice of collecting all exercises in a single section avoids distracting the readers with practical details that are not required for understanding the circuit's working principles.

Appendix A gives a very simple treatment of the transistor and Appendix B is a concise collection of math tools, that are frequently used in the rest of the book, and that are briefly explained for the less expert reader. Appendix C and D give details on the commercially available passive and active components, useful for practical purposes.

Sometimes references to data available in the Web are suggested, mostly to Wikipedia.

1. Introduction

This short chapter is devoted to those who never studied electronic circuits, and it may be skipped by anyone who yet knows what is a network made of current and voltage sources, resistor, capacitors and inductances[1].

1.1. Voltage and current signals

Any physical quantity may be used to transfer information, i.e. as a signal. A signal may be either *analogic* or *digital*. In the first case one has a smooth change of the physical quantity, in the time-domain, in the second case the quantity may take only discrete values (usually two): e.g. ZERO value (also named NOT, or OFF, or LOW), and ONE value (YES, or ON, or HIGH).

In electronics two signal are taken into consideration: *voltage* (V) and current (I).

Voltage (unit: volt=V) is a measurement of the electric potential difference between two points in a circuit; current (unit: ampere=A) is a measurement of the charge carriers flux per time unit from one point of a circuit to another one.

The charge carriers (unit q=Coulomb) are conventionally assumed to be *positive*, moving from point at higher potential to points at lower potential. In the real world they may be either positive (holes in semiconductors) or negative (electrons in metals and semiconductors)

1.2. Resistors, capacitors, inductances, signal sources

Resistors are bipolar passive elements, made of conductors connecting two points (A and B) in a circuit. The voltage V_{AB} at the resistor's ends (=potential difference between V_A and V_B) and the current I flowing across the resistor are bound by a linear relation (the Ohm's Law) $V_{AB}=RI$, where R is a positive constant, that measure the electrical resistance (unit: ohm, symbol Ω). The resistance of a homogeneous cylindrical conductor is given, in terms of the material resistivity ρ by the equation $R=\rho l/S$, where *l* is the conductor length and S the cross-section.

A finite electrical resistance is associated to any conductor; but the copper wires connecting various elements in a circuit, due to the low copper resistivity, are normally assumed to have zero resistance.

Capacitors are bipolar passive elements, made of two electrodes separated by a dielectric layer; the voltage V_c across the capacitor's ends obeys the equation $V_c=q/C$, where q is the charge[2] accumulated at the electrodes and C is a constant named capacity (unit: farad= F).

---

[1] For a more detailed introduction: *Electricity* by A. Shure, or *Electronic Circuits and Applications* by S. Senturia eand B.Wedlock (Chapt. 2); se also http://en.wikipedia.org/wiki/Network_analysis_%28electrical_circuits%29

[2] The charge q has opposite sign and equal values on the two electrodes. Capacitors may be of different types: see app C.3.

The wires connecting various elements in a circuit may also be seen as electrodes separated by dielectric medium (air), so that they form capacitors distributed in the whole circuit. But the small value of these *parasitic* capacitances makes them negligible in most cases. The current $I = \partial q/\partial t$ flowing from one electrode to the other one, may be written $I = C\partial V_c/\partial t$.

Inductors are bipolar passive elements, made of a conductor wound into a coil; the voltage V_L across the inductor's ends is proportional to the flowing current: $V_L = L\partial I/\partial t$. The constant L is the inductance (unit: henry = H)[3], which measures the efficacy of the inductor in changing the linked magnetic field when a current flows across it.

The symbols representing resistors, capacitors and inductors are given in figure 1.1. Details on different types of these elements are reported in Appendix C.

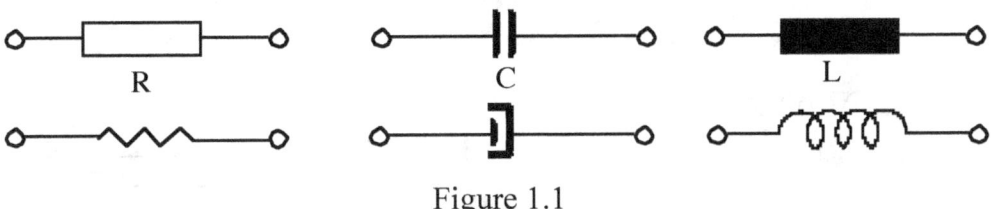

Figure 1.1

An *ideal* voltage source is an *active* bipolar device, generating a potential difference between its two poles ($V_{AB} = V_0$, also named electromotive force), which does not depend on the current flowing across it. A *real* voltage source (constant: battery, or variable: oscillator, pulser, electrical noise ...) always includes an electric resistance R_i, named *internal resistance* of the source: $V_{AB} = V_0 - R_i I$. Similarly, an *ideal current source* is an active bipolar device, generating a current which does not depend on the voltage across its terminals.

1.3. Linearity, superposition, Kirchhoff's Laws

A network is said to be *linear* if in each branch a linear relation[4] holds between voltage and current. Ideal resistors, capacitors and inductors are linear elements.

Any linear network obeys the *superposition principle*[5]. This principle states that the *net response* at a given place and time caused by two or more sources is the *sum* of the responses which would have been caused by each source individually (i.e. by switching off all the other sources, which means replacing all other voltage sources by a short circuit, and all other current sources by an open circuit).

---

[3] The physical meaning of inductance may be deduced from Faraday's Law which states that the electromotive force (EMF) induced into any closed circuit is equal to the time rate of change of the magnetic flux through the circuit. (see http://en.wikipedia.org/wiki/Faraday%27s_law_of_induction) A useful mechanical analogy is obtained by substituting the electric current with speed, the induced EMF with inertial force, and the inductance with mass.

[4] A function f is *linear* if for any two inputs x_1 and x_2 $f(x_1 + x_2) = f(x_1) + f(x_2)$.

[5] See http://en.wikipedia.org/wiki/Superposition_principle

The following rules hold, named *Kirchhoff's Laws* in any linear network:

1) the algebraic sum of all *voltages* in any single *loop* (or *mesh*)[6] is zero;

2) the algebraic sum of all *currents* entering a single *node* is zero.

The first Law is named *Kirchhoff Voltage Law* (KVL), the second one *Kirchhoff Current Law* (KCL). Using these rules and the Ohm's Law, solving any linear system becomes quite easy: e.g. becomes immediate calculating the equivalence of various combinations of resistors, capacitors and inductors (see figure 1.2).

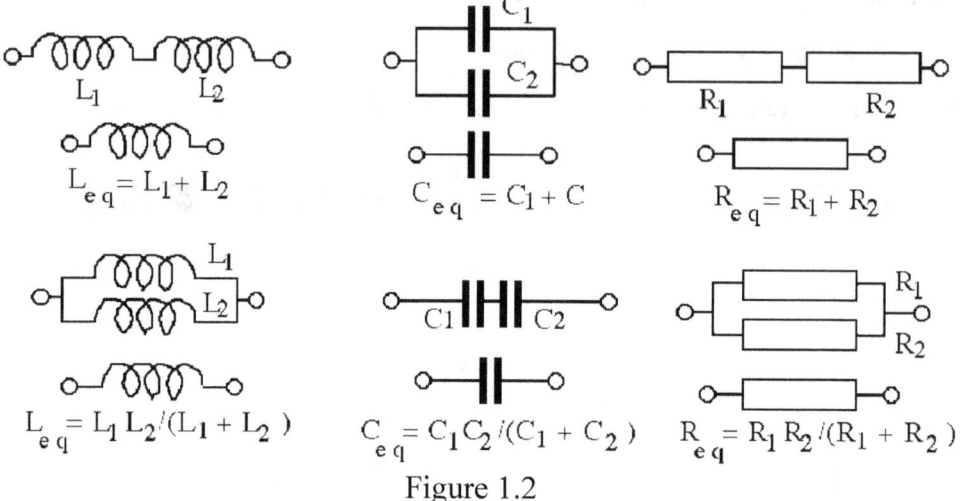

Figure 1.2

Two resistors R_1, R_2 [or inductors[7] L_1, L_2] placed in *series* are equivalent to a single resistor R_{eq} [or inductor L_{eq}] whose value is the sum of the two values $R_{eq}=R_1+R_2$ [$L_{eq}=L_1+L_2$]. The resistor R_{eq} [or L_{eq}], equivalent to two resistors R_1, R_2 [or L_1, L_2] in *parallel*, is R_{eq} ($R_1 \| R_2$)=$R_1 R_2/(R_1+R_2)$ [or $L_{eq}=L_1 L_2/(L_1+L_2)$] [8]. The symbol $\|$ is frequently used to indicate the *parallel* combination of two elements.

Two capacitors placed in *parallel* are equivalent to a single capacitor whose value is the sum of the two values C_{eq} ($C_1 \| C_2$)=C_1+C_2, while two capacitors C_1, C_2 in *series* are equivalent to a single capacitor whose value is $C_{eq}=C_1 C_2/(C_1+C_2)$.

A frequent calculation is the subdivision of a voltage by means of two resistors in series as shown in figure 1.3. This simple circuit, where the

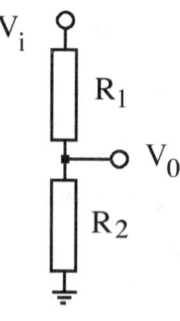

Figure 1.3

---

[6] A node is a point of the network that join two or more branches, a mesh is a closed loop that starting from a node returns to the same node without crossing a brach more than one time.

[7] Here we assume inductors with negligible *mutual inductance*.

[8] These relations hold only if the inductors do not interact, i.e. if the mutual inductance M is negligible; this occurs when the magnetic field linked with one inductor in not linked to the other inductor. Otherwise one must account for M, as in the case of primary and secondary windings in a transformer.

voltages are referred to the common ground, is named *resistive divider*. The same current I flows through the two branches R_1 and R_2. The Ohm's Law gives: $V_i = I(R_1 + R_2)$ and $V_o = IR_2$.

By eliminating I from the previous equations, one gets for the output voltage: $V_o = V_i R_2 / (R_1 + R_2) = \beta V_i$, where β is named *partition fraction* of the input signal V_i.

2. Operational amplifiers

A large part of modern electronic circuits is made of Integrated Circuits (IC), which are composed by many microelements, both active (as transistors) or passive (as resistors, capacitors, inductors…). Among the linear IC most part are operational amplifiers (OA).

Understanding the working principle of OA is possible without entering into the details of their internal structure. They may be considered as *black boxes*, i.e. as objects completely characterized by their functional properties, or by the relations they establish between input and output signals.

2.1. Basic concepts and definitions

The Operational Amplifier[9] (AO) is an integrated circuit, made of resistors, capacitors, diodes and transistors encapsulated into a single small container[10], plastic or metallic, which is normally connected to the rest of circuitry through spring-loaded contacts (Figure 2.1).

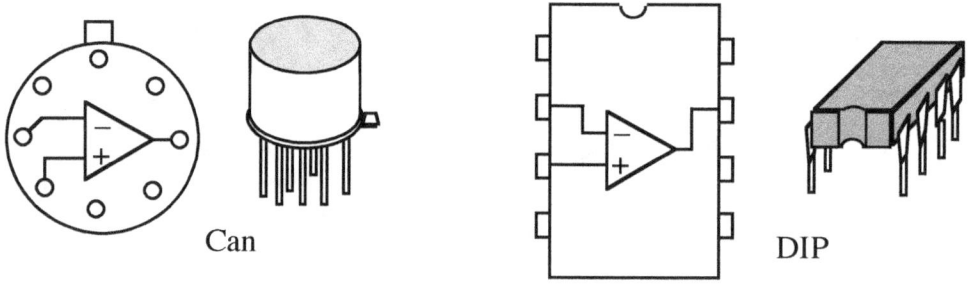

Can   DIP

Figure 2.1

The OA may be functionally defined as *differential amplifier*, i.e. an active device with three ports[11], generating, at the output port, a voltage proportional to the difference between the voltages entering the two input ports. All these voltages are referred to the common potential, named *ground potential*.

The ratio between the output voltage and the input potential difference is named *open loop differential gain* A_d. The value of A_d for d.c. or low frequency signals ($f < f_o \approx 100\,\text{Hz}$) is very high ($A_d \approx 10^5$).

---

[9] The name Operational Amplifier was invented by people who dealt with *analogic* electronic calculators, (see e.g. http://en.wikipedia.org/wiki/Analog_computers These calculators, now superseded by *digital* calculators, used OA in order to process voltage signals executing operations as sum, subtraction, multiplication, division, integration etc.. A simple example is here given in chapt 8.4.

[10] The pinout is generally circular in the metal can models and Dual-In-line Package (DIP) in the plastic models. More details in Appendix D2.

[11] Some rare models offer also offer also differential output.

The graphic symbol commonly used for indicating OA is shown in figure 2.2. Here V_1 and V_2 are the input voltages and V_o is the output voltage, while the symbols (–) and (+) indicate the *inverting* and *non-inverting inputs* (or *channels*), respectively.

The power supply ports (named V_{cc}^+ and V_{cc}^-) in figure 2.2 are frequently omitted in simplified drawings. The voltages supplied to these ports have usually equal values with opposite sign (from ± 5 V to ± 20 V) in *dual* power supply, or typically $V_{cc}^+ = 3\,V \div 30\,V$ and $V_{cc}^- = 0\,V$ in *unipolar* power supply. In the following, where there is no different specification, the default power supply is *dual*.

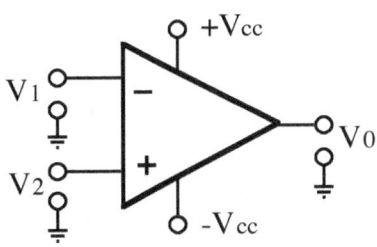

Figure 2.2

The OA amplifies the difference $V_d = V_2 - V_1$ between input voltages only when the device operates in the *linear region*, that is limited by very small values of $|V_2 - V_1|$. This is due to the *finite values* of both open loop differential gain and of power supply voltages.

For larger values of $|V_d|$ the OA *saturates*, which means that its output voltage reaches the limit values V_{cc}^+ or V_{cc}^-, for $V_2 > V_1$ or $V_2 < V_1$, respectively.

The open loop differential gain A_d is the result of superposition of the two channel gains. The signal fed to the inverting input appears at the output amplified by a factor $(-A^-)$ and added to the signal fed to the non-inverting input (which is amplified by a factor A^+). As a result we get:

$$V_0 = -A^- V_1 + A^+ V_2 \qquad [2.1]$$

The absolute value of the gain in the two channels is very similar but not identical, so that normally the open loop differential gain is given as their mean value:

$$A_d = \tfrac{1}{2}\left(|A^+| + |A^-|\right). \qquad [2.2]$$

The absolute value of the difference between the two absolute values is named *common-mode gain*:

$$A_{cm} = \left|\left(|A^+| - |A^-|\right)\right| \qquad [2.3]$$

It is easy to guess that always $A_{cm} \ll A_d$.

The difference between the two input signals is named *differential signal* (note that while the input signals are referred to common ground potential, the differential signal is a *floating* signal):

$$V_d = V_2 - V_1 \qquad [2.4]$$

And their mean value is named *common-mode signal* (referred to ground potential):

$$V_{cm} = \tfrac{1}{2}(V_2 + V_1). \qquad [2.5]$$

From the above definitions follows that the input signals may be written in terms of differential signal and common-mode signals:

$$V_1 = V_{cm} - \tfrac{1}{2} V_d \quad \text{and} \quad V_2 = V_{cm} + \tfrac{1}{2} V_d . \qquad [2.6]$$

The output signal V_o may therefore be written in terms of V_d, V_{cm}, A_d and A_{cm}:

$$V_o = A^+ V_2 - A^- V_1 = A_{cm} V_{cm} + A_d V_d . \qquad [2.7]$$

The ratio, measured in decibel (dB), between A_d and A_{cm} is named *Common-Mode Rejection Ratio (CMRR)*. Typical value for CMRR is 100 dB.

$$\text{CMRR} = 20 \log_{10} (A_d / A_{cm}). \qquad [2.8]$$

Another important parameter that describes the behavior of real OA is V_{os} (*input offset voltage*), i.e. the voltage required across the OA input terminals to drive the output voltage to zero.

V_{os} is normally small (of the order of millivolt), and many OA provide also pins used to zero this offset (*offset null* pins). The value of V_{os} depends on temperature and on power supply: the sensitivity to such parameters is measured as $\partial V_{os}/\partial T$ (V_{os} *temperature coefficient*), typically some $\mu V/K$, and as PSRR (*Power Supply Rejection Ratio* = $\partial V_{cc}/\partial V_{os}$) of the order of 100 dB.

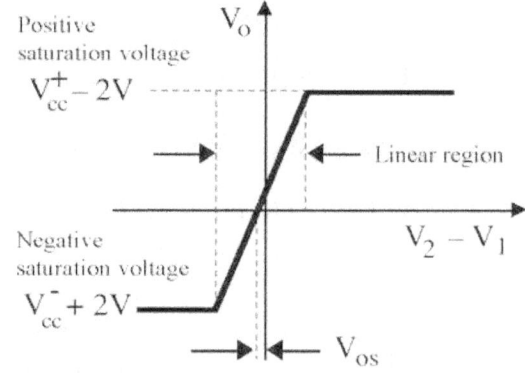

The maximum swing of the output voltage V_o, in linear regime, has normally[12] a value smaller than the power supply value: typically $V_{cc}^- + 2V \leq V_o \leq V_{cc}^+ - 2V$.

Figure 2.3 shows an example of dual power supply with negative V_{os}. In this figure the V_{os} value was exaggerated in order to make it visible. The linear region is defined as the maximum swing of differential input voltage that does not bring the output into saturation.

The *input bias current* I_b may be neglected in a first approximation, being small with respect to other currents normally flowing within the circuit. The OA have high *input impedance*[13] ($Z_{in} \approx 10^6 \div 10^{11}\ \Omega$) and small output impedance ($Z_{out} \approx 1 \div 100\ \Omega$). The input impedance Z_{in} is the ratio between the input voltage and the current injected into the input. The *output impedance* Z_{out} may be seen as the internal resistance of the controlled-voltage-source $V_o = A_d (V_d)$ driven by the

---

[12] In some OA (named RAIL to RAIL) the output voltage swing cover the full power supply range .
[13] More in details, the input impedances $Z_{1,2}$ of single input terminals differs from the differential input impedance Z_d: usually $Z_{1,2} > 10^9\ \Omega$ and $Z_d \approx 10^{-2} Z_{1,2}$.

input differential voltage V_d (see Figure 2.4).

2.2. The Ideal Operational Amplifier

The model of *ideal operational amplifier*, used in simplified analysis, is defined by the following approximations for a *voltage-controlled voltage source*:

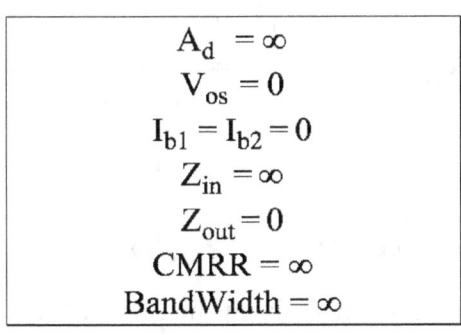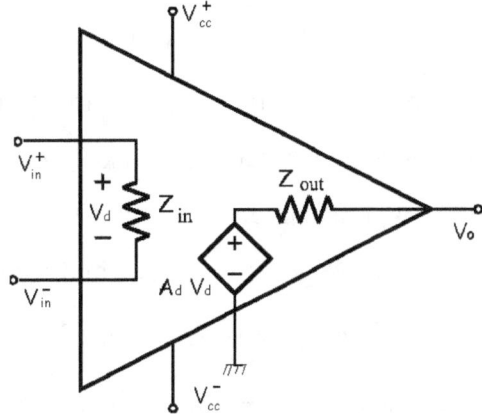

Figure 2.4

By using the approximate model of *Ideal Operational Amplifier* one may reach a faster understanding of complex circuits involving OA. Taking into account the *non-ideal* characteristics of real OA may later refine the analysis.

At first sight the model of ideal OA might appear useless within the linear region, because any finite differential input voltage V_d would produce saturation for $A_d=\infty$. We will however see in the next chapter that, by using some negative feedback (that reduces the differential input voltage), the OA may be always kept within the linear region.

2.3. Real Operational Amplifiers

The following table 1.1 gives a summary of the typical values of essential parameters for different types of commercial OA: input stage made by bipolar junction transistors (BJT) or by field effect transistors (FET) or by metal oxide transistors (MOS).

Input stage	V_{os} (mV)	I_b (pA)	I_{os} (pA)	CMRR (dB)	ω_1 (MHz)
Bipolar	0.01÷2	≈100.000	≈10.000	≈90	1÷2
FET	0.5÷5	5÷30	0.5÷5	≈90	1÷5
MOS	0.1÷0.5	1	0.5	90÷110	1÷2

Table 1.1

The parameter I_{os} (*input offset current*) is the difference between the two input currents: $I_{os}=|I_{b1}|-|I_{b2}|$. Normally I_{os} is smaller than I_b by an order of magnitude: ($I_{os}/I_b \approx 0.1$).

The open loop differential gain A_d will be simply written A from now on.

It is a complex function of the signal frequency f: $A = A(j\omega)$, (where $\omega = 2\pi f$ is the angular frequency and $j = \sqrt{-1}$ is the imaginary unit[14]), that resembles the transfer function of a low-pass filter[15]: $A(j\omega) \approx A_0(1+j\omega/\omega_0)^{-1}$.

In a plot of $\log|A(\omega)|$ versus $\log(\omega)$, the function $|A(\omega)|$ may be approximated by a piecewise linear function; in this case the graph is named Bode plot[16] (figure 2.5).

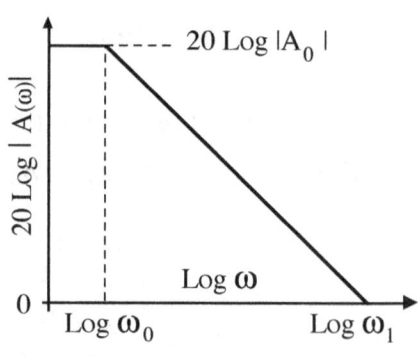

Figure 2.5

In fact for $\omega \ll \omega_0$ we get $|A(\omega)| \approx A_0$ and for $\omega \gg \omega_0$ we get $|A(\omega)| \approx A_0 \omega_0/\omega$.

The parameter $f_0 = 2\pi/\omega_0$ is named *break frequency*, and it is normally of the order of few Hz.

The product $A_0\omega_0$, where $A_0 = A(0)$ is the value of gain at zero frequency, is named *gain-bandwidth product* (GBP or GBWP).

The frequency ω_1 at which the open loop gain is 1 is named *unity-gain frequency* and its value measures the OA speed. In the Bode approximation we get $\omega_1 = A_0\omega_0 = GBP$, and ω_1 is the intercept on the horizontal axis: in fact for $A(\omega_1) = 1$ we obtain $20\log[A(\omega_1)] = 0$.

The maximum current (I_{OAmax}) that a common OA may supply to the output shorted to ground is of the order of few mA, but there are also models with a power output buffer providing currents up to a few ampere[17].

Two parameters closely related to GBP are: the τ (*rise time*), which is the time required to bring the output voltage from 10% up to 90% of the steady value when we fed to the input a step signal, and the *slew rate*, which is the maximum speed of the output voltage changes (usually measured in V/μs). The rise time depends on the closed loop gain G, and practically is reciprocal of the bandwidth: $\tau \approx 1/\Delta\omega = G/\omega_1$. The slew rate is generally measured with G=1, and it is limited by I_{OAmax}.

---

[14] For some details on complex notation and imaginary unit see Appendix B
[15] For details on filters see chapter 8.
[16] See: http://en.wikipedia.org/wiki/Bode_plot
[17] See for example National μA759 and μA791, Siemens TC365, SGS L165, Burr-Brown 3571, ...

3. The operational amplifier as signal processor

By providing the OA with *negative feedback*, using passive elements as resistors or capacitors, we obtain an amplifier that has lower gain but much higher stability. Negative feedback means that a fraction of the output voltage is fed to the inverting input of the OA. This type of configuration is named *closed loop*, and we'll use the symbol G to indicate the closed loop gain, to distinguish it from the open loop gain A: G<<A.

With negative feedback, the circuit's overall gain and other parameters become determined more by the feedback network than by the op-amp itself. If the feedback network is made of components with relatively constant, stable values, the unpredictability and inconstancy of the OA parameters do not seriously affect the circuit's performance.

Using negative feedback we may build circuits that perform on voltage signals operations as sum, subtraction, differentiation, integration.

When the OA operates outside the linear region, we may use both negative and positive feedback to obtain switching circuits (threshold detectors, timers, pulsers ...).

With positive feedback within linear region we may also build oscillators, phase shifters ...

Most of the devices containing OA may be easily analyzed by using the ideal AO approximation, the Kirchhoff's Laws and the *superposition principle*.

In this chapter we'll study the basic configurations: inverting and non-inverting amplifier, the summer and the differential amplifier. We'' also investigate the effects of finite open loop gain ($A \neq \infty$) and of finite bias currents ($I_b \neq 0$).

3.1. Inverting amplifier

The inverting amplifier circuit is shown in in Figure 3.1.
We use hereafter the ideal AO approximation.
From $I_{b2}=0$, we get $V_2=-RI_{b2}=0$. The role of resistor R, whose value is here arbitrary, will be clear when we'll take into account real OA with $I_b \neq 0$.
Assuming infinite value for the open loop gain ($A=\infty$), the input differential voltage must be zero.
In fact $V_d=V_2-V_1=V_o/A=0$, i.e. $V_1=V_2$. Moreover, because $I_{b2}=0$ we get $V_2=0$, so that the feedback keeps both the non-inverting and the inverting

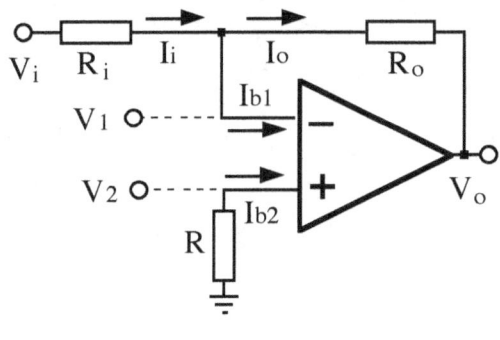

Figure 3.1

terminal bound to a *virtual ground*[18].

Because also $I_{b1}=0$, we get $I_o = I_i - I_{b1} = I_i$, and through the Ohm's Law (I=V/R) we obtain the relation $(V_i-V_1)/R_i=(V_1-V_o)/R_o$, i.e. $V_i/R_i=V_o/R_o$, that gives the *closed loop gain* $G=V_o/V_i$:

$$G = V_o/V_i = -R_o/R_i . \qquad [3.1]$$

Relation 3.1 shows that the closed loop gain depends only of the values of R_o and R_i.

3.2. Non-inverting amplifier

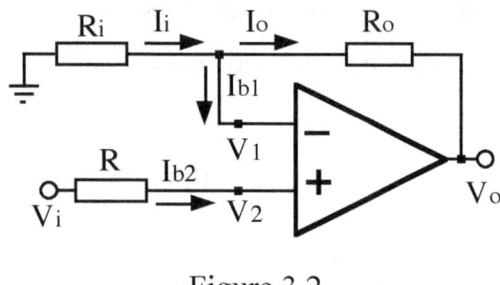

Figure 3.2

Another basic configuration is the non-inverting amplifier, shown in Figure 3.2.

The ideal OA approximation gives again $V_2=V_i$ because $I_{b2}=0$, (no voltage across R).

Also here $V_2=V_1$, because $A=\infty$, and therefore $V_1=V_i$. Again $I_0 = I_i - I_{b1} = I_i$ and through the Ohm's Law (I=V/R) we obtain $I_i=-V_1/R_i=I_0=(V_1-V_o)/R_o$.

Replacing V_1 b V_i in the last relation we get *closed loop gain* $G=V_o/V_i$:

$$G= V_o/V_i = 1+R_o/R_i \qquad [3.2]$$

Again the value of G depends only on the values of the resistors in the feedback network.

3.3. Voltage follower

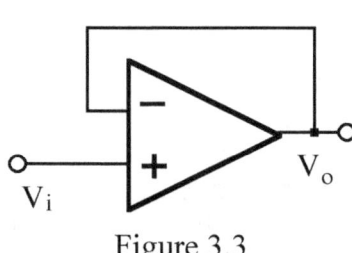

Figure 3.3

A particular type of non-inverting amplifier is obtained for $R_i =\infty$ (open circuit): for any value of R_o (e.g. for $R_o=0$, as in Figure 3.3) we get $G=1$.

This circuit is named *buffer* (or *voltage follower* because $V_o=V_i$).: it offers high input impedance and low output impedance; it therefore does not load the input signal and approximate at the output an ideal voltage source.

---

[18] A node in a circuit is named *virtual ground* when it is bound to ground potential *without a physical short circuit* to ground.

3.4. The differential amplifier

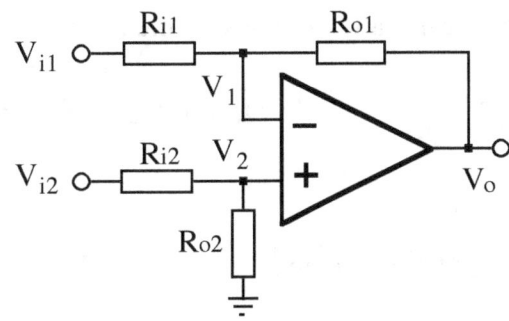

Figure 3.4

Figure 3.4 shows the circuit of a differential amplifier with negative feedback. It may be seen as the superposition of an inverting and a non-inverting amplifier.

The output voltage is the sum $(V_{o1}+V_{o2})$ of two contributions: the one due to the inverting amplifier (when we switch-off the source V_{i2}): $V_{o1}=-(R_{o1}/R_{i1})V_{i1}$, and the one due to the non-inverting amplifier (when we switch-off the source V_{i1}): $V_{o2}=(1+R_{o1}/R_{i1})V_{i2}R_{o2}/(R_{o2}+R_{i2})$.

In the particular case of *balanced* amplifier[19]: $R_{i1}=R_{i2}=R_i$ e $R_{o1}=R_{o2}=R_o$, we obtain at the output: $V_o=V_{o1}+V_{o2}=(R_o/R_i)(V_{i2}-V_{i1})$.

The closed loop gain is therefore:

$$G_d = V_o / V_d = R_o / R_i \qquad [3.3]$$

In the more complex case of *unbalanced* amplifier ($R_{o1} \neq R_{o2}$ and/or $R_{i1} \neq R_{i2}$) the analysis is made easier if we write V_o in terms of V_{cm} and V_d:

$$V_0 = \frac{R_{o2}R_{i1} - R_{o1}R_{i2}}{R_{i1}(R_{i2}+R_{o2})} V_{cm} + \frac{R_{o1}}{R_{i1}} \left[\frac{1+R_{i1}/R_{o1}}{1+R_{i2}/R_{o2}} \right] \frac{1}{2} V_d = G_{cm} V_{cm} + G_d V_d \qquad [3.4]$$

relation that may be deduced from the circuit of Figure 3.5, using the superposition principle.

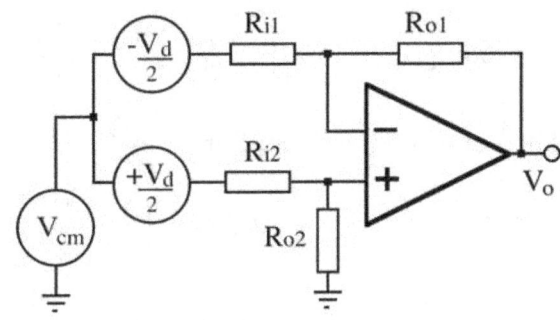

Figure 3.5

In the special case of *equal ratios* $R_{o1}/R_{i1} = R_{o2}/R_{i2} (= R_o/R_i)$, eq. 3.4 shows that the common-mode gain is $G_{cm}=0$, and the differential gain is $G_d=R_o/R_i$. This demonstrates (for $V_{os}=0$) that we may *balance* the differential amplifier simply adjusting one of the four resistors.

The effect produced by a small unbalance may be evaluated letting $R_{i1}=(1+x)R_i$, $R_{i2}=(1-x)R_i$, $R_{o1}=(1-x)R_o$, $R_{o2}=(1+x)R_o$: substituting into eq. 3.4 yields $G_{cm} \approx 4x(R_o/R_i)/(1+R_o/R_i)$, that for $R_o \gg R_i$ gives $G_{cm} \approx 4x$. Using precision resistor (x=1%) in the worst case we get $G_{cm}=0.04$.

---

[19] To balance a differential amplifier means to minimize the common-mode gain (i.e to maximize CMRR). We must remember that sole role is played by the output impedances of the sources V_{i1} e V_{i2}, that add up to R_{i1} and R_{i2}, respectively.

3.5. Inverting summer

We may easily add voltages by means of the circuit shown in Figure 3.6.

Because $I_1 + I_2 + \ldots I_n = I_o$, we get $V_0 = -R_o \left(\dfrac{V_1}{R_1} + \dfrac{V_2}{R_2} + \ldots \dfrac{V_n}{R_n} \right)$. With $R_o = R_1 = \ldots R_n$, more simply $V_o = -\Sigma_i V_i$.

The output voltage may be written $V_o = -R_o \Sigma_i I_i$, where I_i are the currents injected into the inverting node.

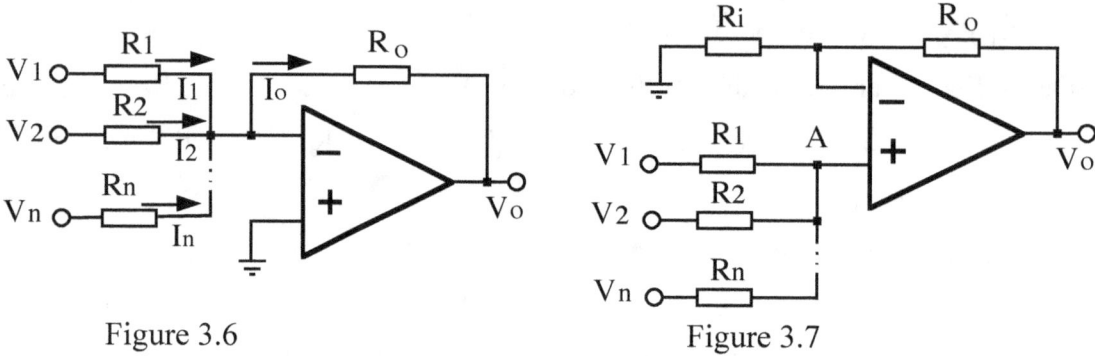

Figure 3.6   Figure 3.7

3.6. Non-inverting summer

In the non-inverting configuration as in Figure 3.7, the analysis is equally simple. At the node A the sum $\Sigma_i I_i$ of currents I_i is zero, in fact $I_b = 0$; by writing $\Sigma_i I_i$ in terms of Ohm's Law (I=V/I):

$$\frac{V_1 - V_A}{R_1} + \frac{V_2 - V_A}{R_2} + \ldots + \frac{V_n - V_A}{R_n} = 0$$

or, letting $R^* = R_1 \| R_2 \| \ldots \| R_n$:

$$\frac{V_1}{R_1} + \frac{V_2}{R_2} + \ldots + \frac{V_n}{R_n} = V_A \left(\frac{1}{R_1} + \frac{1}{R_2} + \ldots + \frac{1}{R_n} \right) = \frac{V_A}{R^*}.$$

The non-inverting amplifier gives $V_o = G V_A$ with $G = 1 + R_o / R_i$ and finally we obtain:

$$V_0 = G R^* \left(\frac{V_1}{R_1} + \frac{V_2}{R_2} + \ldots + \frac{V_n}{R_n} \right) = G R^* \Sigma_i (V_i / R_i) = \Sigma_i \alpha_i V_i$$

where ($\alpha_i = R^*/R_i$). In other word the output voltage is a linear combination of the input voltages.

If the resistors $R_1 \ldots R_n$ are all equal we get $V_o = (G/n) \Sigma_i V_i$, which states that the output voltage is the mean value of input values.

For $G = n$, i.e. $R_o = (n-1)R_i$ we finally get simply a *non-inverting summer* of the input voltages.

3.7. Effects of bias currents and offset voltage

In order to control the reliability of the *ideal OA model*, let us now calculate the effects of non-zero bias currents I_b and non-zero input offset voltage V_{os}, while maintaining the approximation $A = \infty$.

The output voltage of ideal OA with zero input voltage should be zero for both inverting and non-inverting amplifier (note that for $V_i = 0$ the circuits of Figure 3.1 and Figure 3.2 are identical. In Figure 3.8 shows the input offset voltage V_{os} fed to the non-inverting terminal: this is equivalent to feed an offset of opposite sign to the inverting terminal.

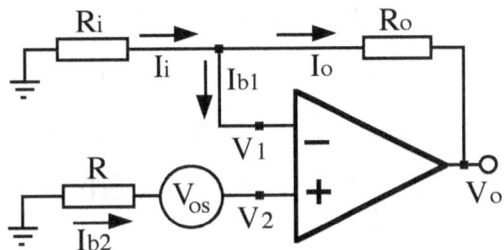

Figure 3.8

So the output voltage in the circuit of Figure 3.8 is the result of $I_b \neq 0$ and $V_{os} \neq 0$ for both inverting and non-inverting amplifier.

Assuming $A = \infty$, we get $V_2 - V_1 = V_o/A = 0$, and therefore $V_2 = V_1$. But now $V_2 = -I_{b2}R + V_{os}$. The relation $I_i = I_o + I_{b1}$ may be written: $-V_1/R_i = (V_1 - V_o)/R_o + I_{b1}$. Eliminating V_1 and V_2, we get: $V_o = V_{os}(1 + R_o/R_i) + R_o I_{b1} - R(1 + R_o/R_i)I_{b2}$.

Defining $I_{os} = (I_{b2} - I_{b1})$, and eliminating I_{b1}, we obtain the relation:

$$V_o = V_{os}(1 + R_o/R_i) - R_o I_{os} + [R_o - R(1 + R_o/R_i)]I_{b2} \qquad [3.5]$$

where $I_{b2} \approx I_{b1} \gg I_{os}$, the input offset voltage V_{os} is amplified by a factor $(1 + R_o/R_i)$.

A proper choice of the resistor R cancels the third term: for $R = R_o R_i/(R_o + R_i)$, i.e. $R = R_o \parallel R_i$, the effect of bias currents is reduced to $R_o I_{os}$, where the *input offset current* I_{os} is normally 10 times smaller than I_b. This particular choice for R is explained by the fact that it balances the input impedances of inverting and non-inverting inputs.

3.8 Effect of the finite open loop gain

Let us investigate now the effect of the *finite value of the open loop gain* for the two basic circuits, maintaining the ideal approximations of symmetric channels ($A_d = A^+ = A^- = A$, i.e. CMRR $= \infty$, or $A_{cm} = 0$), and neglecting bias currents ($I_b = 0$).

The output voltage is again:

$$V_o = A(V_2 - V_1). \qquad [3.6]$$

We'll use a simplified notation by introducing the parameter β (*feedback fraction*), which is the fraction of the output signal that is fed to the input terminals. In previous circuits, the feedback is fed to the inverting terminal (negative feedback), i.e. $\beta = V_1/V_o$, or $\beta = R_i/(R_o + R_i)$.

Let us begin with the inverting amplifier circuit.

We apply the superposition principle to calculate, at the inverting input, the separate contributions of the two sources V_i and V_o, that we rename V_{1i} and V_{1o}.

The signal V_1 at the inverting input may be written as $V_1 = V_{1i} + V_{1o}$ with $V_{1i} = (1-\beta) V_i$, and $V_{1o} = \beta V_o$. We thus obtain $V_1 = V_{1i} + V_{1o} = (1-\beta) V_i + \beta V_o$.

Neglecting I_b we get $V_2 = 0$. And substituting these values V_1 and V_2 into relation [3.6], we obtain:

$$V_o = (1 - 1/\beta) V_i / \{1 + 1/A\beta\} = (-R_o/R_i) V_i / \{1 + 1/A\beta\}. \qquad [3.7]$$

The *closed-loop gain* $G = V_o/V_i$ becomes:

$$G/\{1 + 1/A\beta\}, \qquad [3.8]$$

where G is the value calculated in eq. 3.1 (with $A = \infty$), that may be written in terms of β, as

$$G = 1 - 1/\beta. \qquad [3.9]$$

Let us now investigate the non-inverting amplifier case.

Here we have $V_{1o} = \beta V_o$, and $V_{1i} = 0$, i.e. $V_1 = \beta V_o$. Neglecting I_b we get $V_2 = V_i$, and using again relation [3.6], we obtain:

$$V_o = (1/\beta) V_i / \{1 + 1/A\beta\} = (1 + R_o/R_i) V_i / \{1 + 1/A\beta\}, \qquad [3.10]$$

Which shows that also for non-inverting amplifier the closed-loop gain is:

$$G/\{1 + 1/A\beta\}, \qquad [3.11]$$

when G is the value calculated in eq. 3.2 (with $A = \infty$) that may be written now as

$$G = 1/\beta. \qquad [3.12]$$

The product Aβ is named *loop gain*, and its reciprocal 1/{Aβ} is named *loop gain error*, because it measures how much the *real* circuit differs from an equivalent circuit using *ideal* OA.

For $A\beta \gg 1$ the closed-loop gain is the one calculated in the ideal case.

In other words, the ideal OA approximation holds until $|G| \ll A$.

3.9. *Input and output impedances for real OA in closed-loop configurations*

The ideal OA model gives for the inverting amplifier $Z_{in} = R_i$, and for the non-inverting amplifier $Z_{in} = \infty$, while the output impedance is by definition $Z_{out} = 0$.

The input impedance is defined as $Z_{in} = V_i / I_{in}$.

In the ideal inverting amplifier the input current is $I_{in} = (V_i - \varepsilon)/R_i$, where ε is the *closed-loop differential input voltage*, which is zero in the ideal model $\varepsilon = V_o/A = 0$ (because $A = \infty$).

In the ideal non-inverting amplifier the input current is $I_{in} = I_{b2} = 0$, by definition.

A refined approximation for *closed-loop* Z_{in} and Z_{out}, must take into account the *open-loop* parameters of *real* OA : $A \neq \infty$, $Z_o \neq 0$ (typically $Z_o \approx 100\,\Omega$) finite value of Z_{in}.

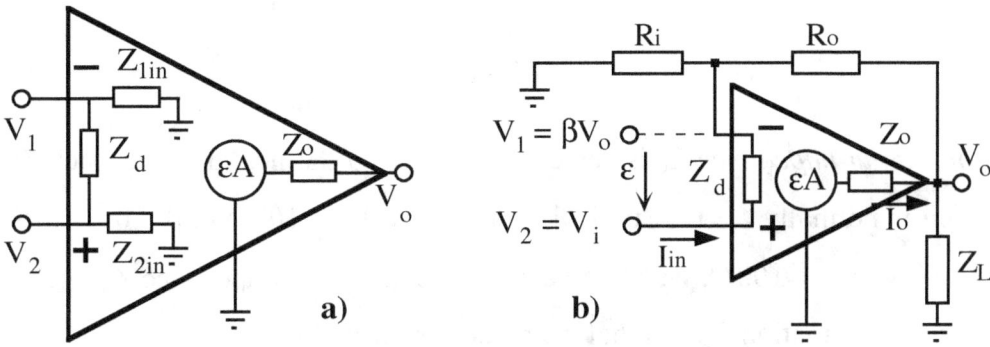

Figure 3.9

Figures 3.9a and 3.9b show models of open-loop and closed-loop *real* OA.

We distinguish between the *impedances of the input channel* $Z_{1,2in}$ and the *differential input impedance* Z_d (usually $Z_{1,2in} > 10^9\,\Omega$ and $Z_d \approx 10^{-2} Z_{1,2in}$). The voltage-controlled-voltage source εA, driven by the differential input voltage $\varepsilon = V_2 - V_1$, is in series with the *open-loop output impedance* Z_o which may be seen as the internal resistance of the source εA.

However, because $Z_d \ll Z_{1,2in}$ we'll neglect $Z_{1,2in}$, assuming $Z_{1in} = Z_{2in} = \infty$.

By definition the *closed-loop output impedance* is $Z_{out} = \partial V_o / \partial I_o$, that may be written here:

$$V_o = G V_i - Z_{out} I_o, \qquad [3.13]$$

equation where the first term at right measures the output voltage with infinite load ($Z_L = \infty$, zero output current), and the second term measures the change of the output voltage due to the current I_o supplied to the load.

From the definition of the *open-loop output impedance* Z_o, seen as internal resistance of the source εA, we get:

$$V_o = \varepsilon A - Z_o I_o. \qquad [3.14]$$

Let us first consider the case of non-inverting amplifier shown in Figure 3.9b.

Recalling that $V_1 = \beta V_o$ and $V_2 = V_i$, where V_i is the voltage fed to the non-inverting terminal (we neglect the voltage drop across the output impedance of the voltage source due to the bias current I_{b2}).

The differential input voltage ε may be written $\varepsilon = V_2 - V_1 = V_i - \beta V_o$, which changes relation

[3.14] into :

$$V_o = \left(\frac{A}{1+\beta A}\right)V_i - \left(\frac{Z_o}{1+\beta A}\right)I_o = \left(\frac{1/\beta}{1+1/A\beta}\right)V_i - \left(\frac{Z_o}{1+\beta A}\right)I_o \qquad [3.15]$$

The first term at right in [3.15] is the same result found in [3.10]→[3.12], i.e. the closed-loop gain $G \approx 1/\beta$ (corrected by the factor $1+1/A\beta$), while the second term, by comparison with [3.13], gives:

$$Z_{out} = \frac{Z_o}{1+\beta A}. \qquad [3.16]$$

The *open-loop output impedance* Z_o is scaled down, due to the negative feedback, by the factor $(1+\beta A)$, which is normally $\gg 1$. A typical example: with $G \approx 100$, i.e. $\beta \approx 10^{-2}$ and with $A \approx 10^5$, i.e. $A\beta \approx 10^3$, and with $Z_o \approx 100\,\Omega$, we obtain a closed loop output impedance $Z_{out} \approx Z_o/A\beta \approx 0.1\,\Omega$. This justify the approximation $Z_{out} \approx 0$ that we made for ideal OA.

To calculate the *closed-loop input impedance* $Z_{in} = V_i/I_{in}$, we note that (assuming $Z_{2in} \gg Z_d$) we may write $I_{in} = \varepsilon/Z_d$, or $Z_{in} = Z_d V_i/\varepsilon$. Now recalling that $\varepsilon = V_i - \beta V_o$, or $V_i/\varepsilon = 1+\beta(V_o/\varepsilon)$ we finally obtain:

$$Z_{in} = Z_d[1+\beta(V_o/\varepsilon)] \approx Z_d(1+\beta A), \qquad [3.17]$$

The approximation shown in 3.17 is due to the voltage drop across Z_o: the voltage εA generated by the voltage-controlled-source, is divided by the Z_o in series with the load Z_L. Therefore $V_o = \varepsilon A Z_L/(Z_o+Z_L)$. Assuming a negligible current[20] in the feedback resistor R_o we obtain:

$$Z_{in} = Z_d[1+\beta A Z_L/(Z_o+Z_L)] \qquad [3.18]$$

Relation 3.18 should replace relation 3.17 when $Z_L \ll Z_o$ (e.g. when output is shorted to ground). Let us evaluate an intermediate case: $Z_d \approx 10\,M\Omega$, $Z_o \approx 100\,\Omega$, $Z_L \approx 100\,\Omega$, $G \approx 1/\beta \approx 100$, $A \approx 10^5$: $Z_{in} \approx Z_d(1+\beta A/2) \approx 501 Z_d \approx 500\,M\Omega$.

We may conclude that the open loop input impedance Z_d is normally multiplied, due to the negative feedback, by a factor of the order of the loop-gain $(\beta A \gg 1)$, justifying the ideal OA approximation $(Z_{in} \approx \infty)$.

A similar analysis may be carried out, with reference to Figure 3.10, for the inverting amplifier.

Here to calculate Z_{out} we still use relations [3.13] and [3.14], but now the differential input voltage is

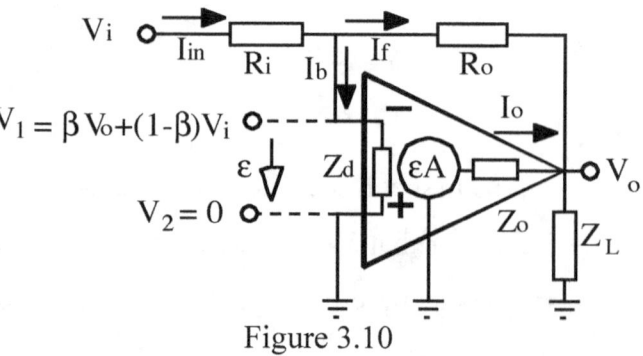

Figure 3.10

---

[20] Accounting for R_o we would get a larger Z_{in}, by a factor close to $1/(1-\beta)$.

$\varepsilon = V_2 - V_1 = -V_1$, because $V_2 = 0$ (to make thing easier we assumed R=0 at the non-inverting input).

Using the superposition principle to calculate $V_1 = V_{1i} + V_{1o}$, we obtain

$$\varepsilon = -V_1 = -[\beta V_o + (1-\beta) V_i], \qquad [3.19]$$

that, with relation [3.14] $V_o = \varepsilon A - Z_o I_o$, gives for V_o

$$V_o = \left(\frac{1 - 1/\beta}{1 + 1/\beta A}\right) V_i - \left(\frac{Z_o}{1 + \beta A}\right) I_o. \qquad [3.20]$$

In relation [3.20] the first term at right is the same result found in [3.7] for the closed-loop gain of non inverting amplifier $G \approx 1 - 1/\beta$, corrected for the *loop gain error*.

Comparing [3.20] with [3.13], we obtain again the expression [3.16] for the closed-loop output impedance $Z_{out} = Z_o/(1+\beta A)$, which is practically zero if $\beta A \gg 1$.

Again we may calculate the closed loop input impedance as $Z_{in} = V_i / I_{in}$, but now the input current I_{in} is the sum of two contributions: the bias current $I_b = -\varepsilon / Z_d$ and the feedback current $I_f = (-\varepsilon - V_o)/R_o$. With the approximation $V_o \approx \varepsilon A$ (again neglecting the voltage drop across Z_o) we obtain $I_{in} \approx -\varepsilon [1/+(1+A)/R_o]$, and using $\varepsilon = -V_1$,

$$I_{in} \approx A V_1 / R_o \qquad [3.21]$$

The Ohm's Law gives also $V_i = V_1 + R_i I_{in}$, or, using [3.21] $V_i \approx (R_o/A + R_i) I_{in}$, which yields:

$$Z_{in} = V_i / I_{in} \approx R_i + R_o/A \approx R_i. \qquad [3.22]$$

The closed-loop input impedance of the real inverting amplifier therefore approximates R_i, the same predicted by the ideal OA model where both input terminals are bound to virtual ground.

4. Some examples

In this chapter we'll study some examples of differential amplifiers with variable-gain, and one amplifier, which may switch from inverting to non-inverting configuration by simply changing the value of two resistors. These circuits may be used as *building blocks* when drawing more complex systems, as we'll see in the next chapters.

4.1. Differential with variable-gain

To change the gain of the differential amplifier of Figure 3.4 while keeping it *balanced*, we should properly adjust two resistors, e.g. by using mechanically-coupled potentiometers.

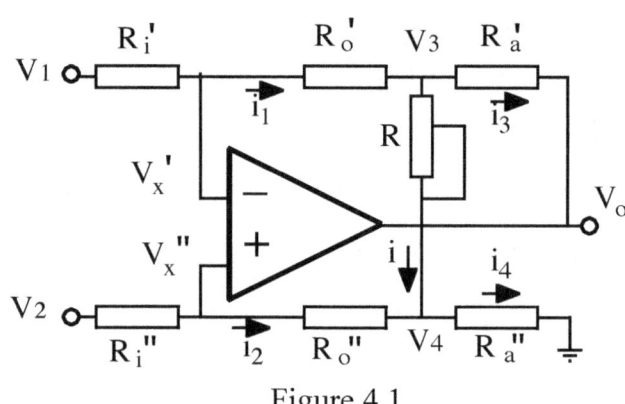

Figure 4.1

This difficulty may be by-passed using the circuit of Figure 4.1 where the gain is adjusted by a *single potentiometer*, after it has been balanced by a proper choice of the six resistors (R' and R"), i.e. letting $R_i'=R_i''=R_i$, $R_o'=R_o''=R_o$ and $R_a'=R_a''=R_a$.

Using the ideal OA approximation $I_b \approx 0$, so that we may write the two equations:

$$(V_1-V_x)/R_i=(V_x-V_3)/R_o \quad \text{and} \quad (V_2-V_x)/R_i=(V_x-V_4)/R_o, \quad [4.1]$$

where we assumed $V_x'=V_x''=V_x$ because $A \approx \infty$.

By subtracting the second equation from the first one we obtain:

$$(V_2-V_1)/R_i=(V_3-V_4)/R_o. \quad [4.2]$$

The ohmic value of the potentiometer R may be written $R = x R_o$, and using the *Kirchhoff Current Law* at the nodes V_3 ($i_1-i=i_3$) and V_4 ($i_2+i=i_4$), we obtain the two equations:

$$(V_x-V_3)/R_o-(V_3-V_4)/xR_o=(V_3-V_o)/R_a \quad [4.3]$$

$$(V_x-V_4)/R_o+(V_3-V_4)/xR_o=V_4/R_a. \quad [4.4]$$

By subtracting eq. [4.3] from eq. [4.4] we get:

$$V_o/R_a = (1+R_o/R_a+2/x)(V_3-V_4)/R_o. \quad [4.5]$$

Finally, by substituting the quantity $V_3-V_4 = (V_2-V_1) R_o/R_i$ taken from [4.2] into [4.5], we obtain the *closed-loop differential gain* $G_d = V_o/(V_2-V_1)$:

$$V_o = (1+R_o/R_a+2/x)(R_a/R_i)(V_2-V_1). \quad [4.6]$$

In the simplest case $R_a = R_o$ the variable gain is :

$$G_d = 2(1+1/x)(R_o/R_1) \ . \qquad [4.7]$$

4.2. Differential with linear variable-gain

In the previous example the gain is a *non-linear function* of R. Using two OA we may obtain a *linear adjustment* of the differential gain. Hereafter we describe three possible configurations. The circuit of Figure 4.2 may be easily understood by considering OA2 as inverting amplifier for the source V_o. The feedback is provided by the potentiometer R with feedback factor $\beta = x/(1-x)$. The gain is therefore $G_2 = -1/\beta = -(1-x)/x = 1-1/x$. To avoid the saturation of OA2 for $x \to 0$, we should place a resistor R' in series to xR. Here to make simpler the analysis we'll neglect R', keeping in mind that x-value must have a lower limit.

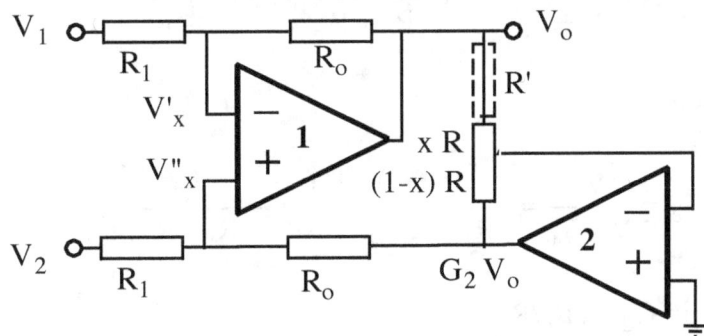

Figure 4.2

Using superposition principle with V_1, V_2 and $G_2 V_o$ sources, we may write:

$$V_o = -\frac{R_0}{R_1} V_1 + \frac{R_0}{R_1 + R_0}\left(1 + \frac{R_0}{R_1}\right) V_2 + \frac{R_1}{R_1 + R_0}\left(1 + \frac{R_0}{R_1}\right) G_2 V_o \qquad [4.8]$$

Solving with respect V_o, we get

$$V_o = (R_o/R_1)(V_2 - V_1)/(1 - G_2) = x\ (R_o/R_1)(V_2 - V_1) \qquad [4.9]$$

So that the differential gain is linear:

$$G_d = V_o/(V_2 - V_1) = x\ R_o/R_1 \qquad [4.10]$$

The maximum gain is for $x = 1$, i.e. for $G_2 = 0$: in this case this circuit is equivalent to that of Figure 3.4. Note that two pair of resistors (R_o and R_1) need to me matched.

A similar configuration is show in Figure 4.3, where the potentiometer is replaced by a resistive divider made of a variable resistor xR and a fixed resistor R; moreover no feedback is fed to the inverting input of OA1, but a negative feedback is provided by OA2, still acting as inverting amplifier with gain $G_2 = -1/x$. Also here a resistor R' should limit the value of G_2 to avoid saturation of OA2.

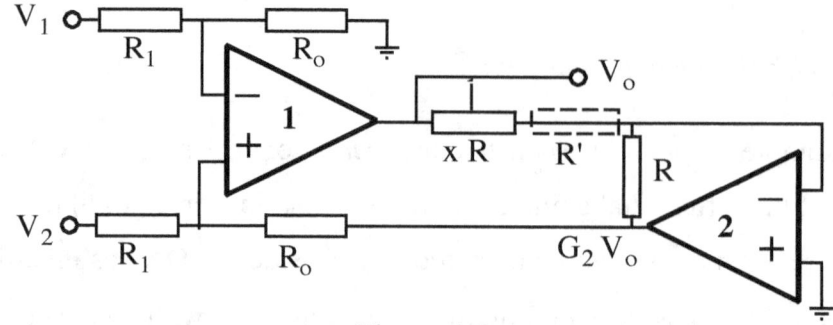

Figure 4.3

Using superposition principle with V_1, V_2 and $G_2 V_o$ sources, we now use the open-loop gain A, that we assume $A \to \infty$.

$$V_o = -AV_1 \frac{R_o}{R_1 + R_o} + AV_2 \frac{R_o}{R_1 + R_o} + AG_2 V_o \frac{R_1}{R_1 + R_o} \quad . \qquad [4.11]$$

Solving for V_o we obtain:

$$G_d = \frac{V_o}{V_2 - V_1} = \left(\frac{R_o}{R_1 + R_o} \right) / \left[\frac{1}{A} + \frac{R_1}{x(R_1 + R_o)} \right], \qquad [4.12]$$

that for $A = \infty$ gives again $G_d = x R_o / R_1$.

Also here two pair of resistors (R_o and R_1) need to me matched.

In the third configuration (figure 4.4) OA1 is a an *inverter* (inverting amplifier with $G = -1$) and OA2 is an inverting summer with $G_2 = -xR/R_1$ for the two sources V_1 and $-V_2$.

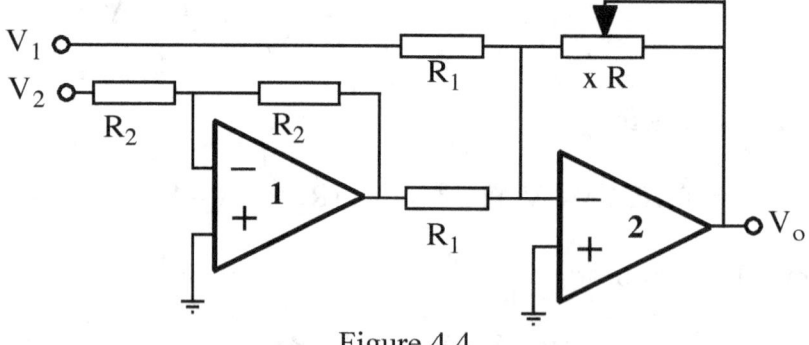

Figure 4.4

Also here two pair of resistors need to me matched (R_1 and R_2): moreover, a good choice would be $R_1 = R_2$ in order to balance the input impedances of the two channels.

4.3. Differential amplifier with variable –gain and high Z_{in}

In the basic differential amplifier of Figure 3.4 (but also in the circuits of Figures 4.1, 4.2, 4.3, 4.4) the input impedance cannot be made very high, in order to avoid vanishing [21] feedback currents.

If the sources feeding the inputs of the basic differential amplifier have large output impedances $Z_{out1,2}$ in relation [3.4] we must replace R_{i1} by $R_{i1} + Z_{out1}$, and R_{i2} by $R_{i2} + Z_{out2}$. This will affect mainly the value of G_{cm}. To avoid this inconvenience we may use the circuit shown in Figure 4.5.

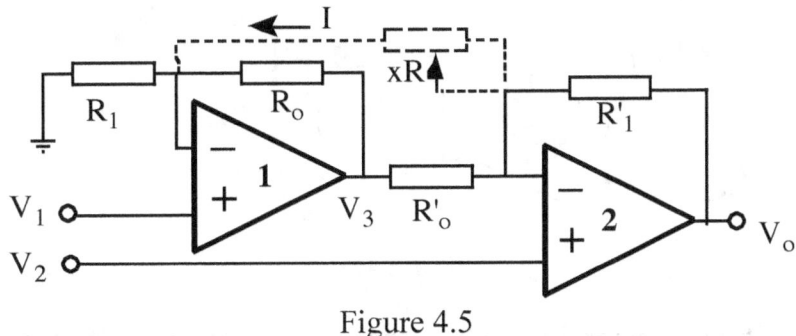

Figure 4.5

To make easier the analysis of this circuit we may start by studying first a simpler one: that obtained removing the branch drawn as dotted line in Figure 4.5. Deleting the potentiometer xR, OA1 behaves as non-inverting amplifier, yielding : $V_3 = (1+R_o/R_1)V_1$.

On the other hand OA2 behaves as inverting amplifier for the source V_3 and as non-inverting amplifier for the source V_2, yielding : $V_o = (1+R'_1/R'_o)V_2 - (R'_1/R'_o)V_3$. Letting $R'_1 = R_1$, $R'_o = R_o$, we get $V_o = (1+R_1/R_o)(V_2-V_1)$, that is a *differential amplifier* with fixed gain $G_d = (1+R_1/R_o)$.

In order to make variable the gain, we insert the potentiometer xR which injects the current $I = (V_2-V_1)/xR$ into OA1 and the current $-I$ into OA2. This current adds a voltage $-IR_o$ at the OA1 output, which is amplified of a factor $(-R'_1/R'_o)$ by OA2, and it adds a voltage IR'_1 at the OA2 output (as we have already seen in the summer circuit of Figure 3.6).

Putting all together we obtain:

$$V_o = \left(1+\frac{R_1}{R_o}\right)(V_2-V_1) - \frac{V_2-V_1}{xR}R_o\left(\frac{R_1}{R_o}\right) + \frac{V_2-V_1}{xR}R_1 = \left(1+\frac{R_1}{R_o}+2\frac{R_1}{xR}\right)(V_2-V_1), \quad [4.13]$$

that for $R_1 = R_o = R$, gives: $V_o = 2(1+R/xR)(V_2-V_1)$. Note that also here we must obviously set a lower limit to x. Relation [4.13] shows that the gain control is *not linear*. Good matching of four resistors is required ($R'_1 = R_1$, $R'_o = R_o$).

---

[21] The current flowing in the feedback must always be much larger than the bias currents.

4.4. Instrumentation amplifier

A frequently used circuit, shown in Figure 4.6, offers the same advantages of the previous one, i.e. high Z_{in} and variable gain, but now we also get *symmetric channels*.

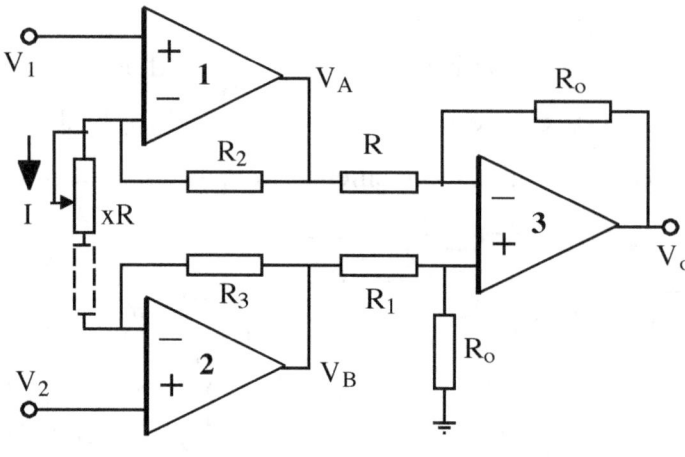

Figure 4.6

We may repeat the trick previously used: without potentiometer xR, the circuit is a basic differential amplifier with a buffer at each input. Therefore the output voltage is

$$V_o = (R_o/R_1)(V_B - V_A), \qquad [4.14]$$

with $V_A = V_1$ and $V_B = V_2$.

The potentiometer xR injects the current $I = (V_2 - V_1)/xR$ into the buffer OA1 and subtracts the same current from the buffer OA2, producing (§3.5) the voltage $-IR_2$ at the output V_A of OA1 and the voltage $+IR_3$ at the output V_B of OA2 :

$$V_A = V_1 - R_2(V_2 - V_1)/xR, \quad V_B = V_2 + R_3(V_2 - V_1)/xR \qquad [4.15]$$

We obtain the same result using the superposition principle, considering the sources V_2 and V_1 first at the output V_A then at the output V_B.

$$V_A = V_1(1 + R_2/xR) - V_2(R_2/xR) \qquad V_B = V_2(1 + R_3/xR) - V_1(R_3/xR).$$

The voltage difference $(V_B - V_A)$ at the input of the basic differential amplifier is:

$$(V_B - V_A) = [1 + (R_2 + R_3)/xR](V_2 - V_1), \qquad [4.16]$$

which gives the output:

$$V_o = (R_o/R_1)(V_B - V_A) = (R_o/R_1)[1 + (R_2 + R_3)/xR](V_2 - V_1) \qquad [4.17]$$

In this circuit, frequently named *instrumentation amplifier*, the gain value is set by a single resistor. Usually $R_2 \approx R_3$, and a resistor in series to xR limits the gain.

4.5. Amplifier with linear gain control from –K to +K

The circuit shown in Figure 4.7a allows *changing the gain from negative to positive values*, by adjusting the potentiometer R

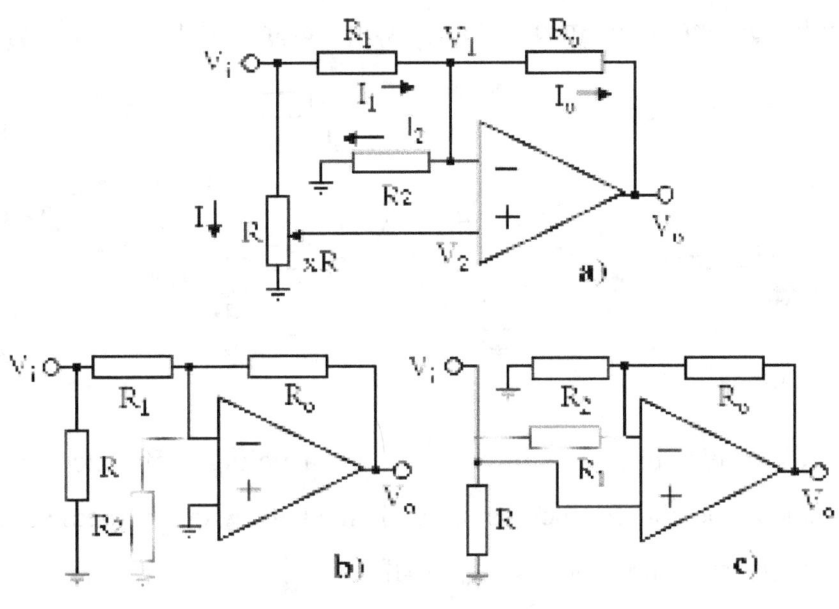

Figure 4.7

This circuits becomes an inverting amplifier with $G=-R_o/R_1$ for $x=0$ (figure 4.7b), and a non-inverting amplifier with $G=1+R_o/R_2$ for $x=1$ (figure 4.7c).

In fact, for $x=0$, V_1 is a virtual ground ($V_2=0$) and resistor R_2 has no effect. For $x=1$, $V_1=V_i$, and resistor R_1 has no effect. Resistor R in both cases loads the source V_i.

In the intermediate cases ($0<x<1$) we simply use the relations: $I_1=I_2+I_o$ and $V_1=V_2=xV_i$.

First relation may be written $(V_i-V_1)/R_1=V_1/R_2+(V_1-V_o)/R_o$, and using the second one we obtain : $V_i(1-x)/R_1 = x V_i/R_2+(x V_i-V_o)/R_o$, that yields the gain G:

$$V_o/V_i = G = R_o/R_1 (x-1) + R_o/R_2 x + x. \qquad [4.18]$$

If we choose the three resistors R_o, R_1, R_2 in order to satisfy the equation $R_o/R_1=1+R_o/R_2=K$ we make the gain to *change linearly from -K to K*.

Relation [4.18] in fact becomes $G=K(2x-1)$.

A particularly useful case is $K=1$, obtained removing R_2 ($R_2=\infty$) and setting $R_o=R_1$. This circuit may be used as multiplier ±1 by switching the non-inverting input between source and ground.

The input impedance varies from R and $R\|R_1$: in fact the input current is the sum $I+I_1$: $Z_i=V_i/(I_1+I)=R R_1/(R_1+R[1-x])$.

5. Reference voltage sources

A *Voltage Reference* (VR) is a source that generates voltage that does not depend on the output current, on temperature and on time (it approximate the ideal voltage source). It must have, therefore, a negligible output impedance and high temperature and time stability.

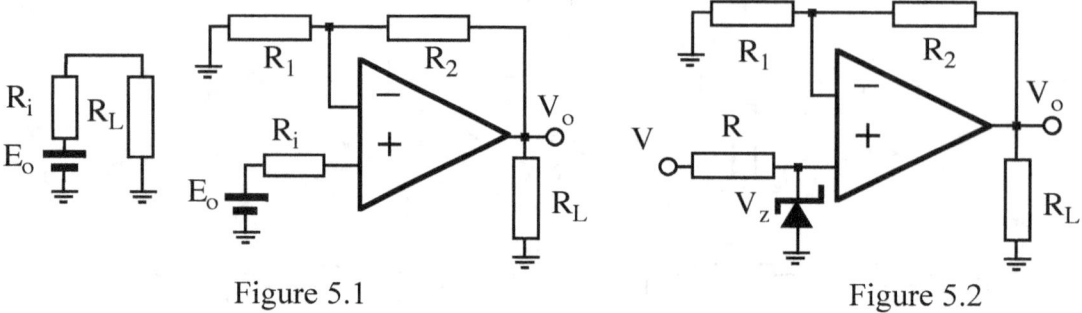

Figure 5.1   Figure 5.2

A battery which has an electromotive force E_o is not a voltage reference because its internal resistance R_i depends on the charge (increasing while the battery discharges) so that the voltage V across the load R_L decreases with time: $V = E_o - R_i I_L$.

A battery followed by an OA, as shown in Figure 5.1, is a better approximation of VR, because the current I_b drained from the battery is small, independent on the load value, and therefore the output voltage $V = E_o - R_i I_b$ well approximates E_o. In the circuit of Figure 5.1 the output voltage V_o may be changed by adjusting the resistors R_1 and R_2. Because $V_o = E_o(1 + R_2/R_1)$, a potentiometer replacing R_2, gives a *linear* voltage regulator.

In the circuit of Figure 5.1 still the output depends on the battery temperature (that affects E_o value). Battery may be replaced by a Zener diode[22] as in the circuit of Figure 5.2, where the unregulated input voltage V may be the OA power supply V_{cc}, and the output becomes: $V_o = V_z(1 + R_2/R_1)$.

Note that V_z depends on the zener current I_z, that depends on V: $I_z = (V - V_z)/R$.

5.1. Voltage sources with zener in the feedback

Because V_z depends (slightly) on the zener bias current I_z, it may be affected by changes in the supply voltage; to avoid this problem we may use the circuit of Figure 5.3, where the zener is part of the feedback loop, which keep constant the bias current I_z.

To analyze this circuit, let us first neglect the divider R_a, R_b and diode D, and we assume that the zener is biased by $V_o > 0$.

---

[22] Details on the zener diode may be found in Appendix A.2. Here it is enough to know that above a threshold value of the inverse bias current, the voltage V_z across a zener diode depends weakly on the current. The value of V_z (named *zener knee voltage*) depends on the type of zener.

Using ideal OA model we may write: $V_1 = V_2$, $V_2 = V_z$, and $V_1 = \beta V_o = V_o R_3/(R_2 + R_3)$. Putting all together we get: $V_o = V_z(1 + R_2/R_3)$. The choice for V_z, R_2 e R_3 must be compatible with the condition $V_z < V_o < V_{cc}$, to avoid saturation.

The zener bias current is $I_z = (V_o - V_z)/R_1 = V_z R_2/R_1 R_3$, i.e. a constant value. The value of R_1 should be maximized (in order to leave most of output current available for the load R_L) but with an upper limit set as $R_1 < V_z R_2 / I_{zmin} R_3$ (to properly bias the zener above the threshold I_{zmin}).

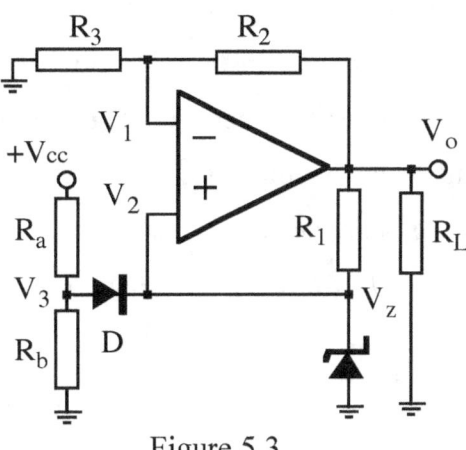

Figure 5.3

The divider (R_a, R_b) is only needed to start the *reverse* current through the zener, and to avoid the second stable state, with a *forward* biased zener ($V_o < 0$) : the voltage V_F across a forward biased zener has in fact a strong dependence on the current. The divider (R_a, R_b) must satisfy the relation $V_3 = V_{cc} R_b / (R_a + R_b) < V_z$, so that, after startup, the diode D is reverse biased.

A similar circuit is shown in Figure 5.4. In this case we get again $V_2 = V_o R_3/(R_2 + R_3)$, but now $V_2 = V_1 = V_o - V_z$ which yields: $V_o = V_z(1 + R_3/R_2)$. The function of the voltage divider (R_a, R_b) and diode D, is the same as for the previous circuit.

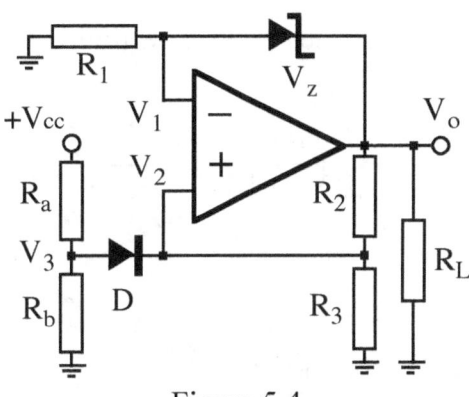

Figure 5.4

5.2. Dual voltage source

A circuit that provides double output voltages ($+V_z$ and $-V_z$), is shown in Figure 5.5. Here the zener is biased through the diode D and resistors R_1, R_3 ; note that the cathode of D is connected by R_3 to $-V_{cc}$ in order to correctly startup the system.

The zener bias current is properly set by R_1: $I_z = (V_z - V_F)/R_1$, where V_F is the forward voltage of D.

The output voltage of OA1 is $+V_z$, , while the output of the inverter OA2 is $-V_z$.

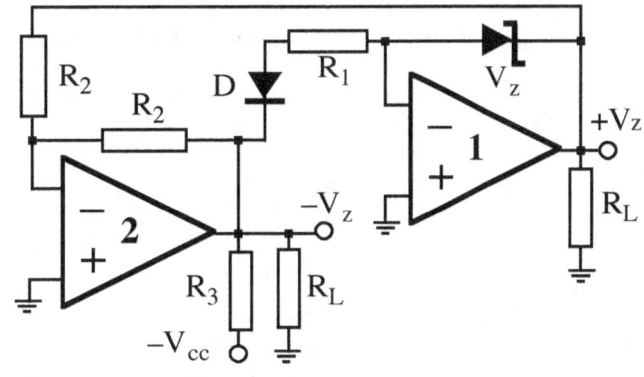

Figure 5.5

6. Voltage-to current converters

If you need to use a resistance thermometer[23] you'll have to measure the voltage across a resistor while keeping constant the current flowing across it. This means that the current fed to the resistor must be independent on the resistor value.

This chapter illustrates some examples of circuits named *voltage to current converters*, or *voltage-controlled current sources,* that supply currents (independent of the load[24]) whose value may be controlled by a voltage source.

6.1. Floating load

If the load can be *floating* (i.e. none of its terminals non physically grounded), the voltage-controlled current-source may be one the two circuits shown in Figure 6.1 .

Here the load R_L is inserted into the feedback loop. In both cases the current $I_L=V_i/R_i$ flowing across the load is controlled by the input voltage V_i and scaled by the resistor R_i. When $V_i>0$, the direction of the current is that marked by the arrow. When $V_i<0$, the output voltage changes sign V_o as well as the current direction.

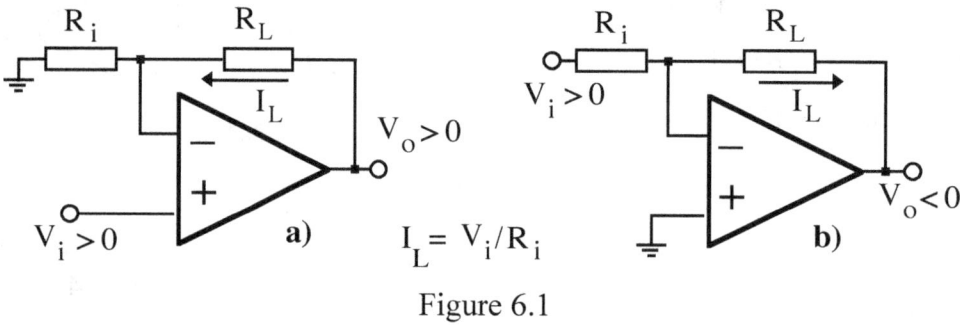

Figure 6.1

In the circuit 6.1a the voltages at the load ends are V_o and V_i, the input impedance is high (the OA closed-loop input impedance Z_{in}) and the current is limited[25] either by the maximum value supplied by the OA, or by the value $I_{max} \approx V_{cc}/(R_i+R_L)$.

In the circuit 6.1b the voltages at the load ends are V_o and virtual ground, the input impedance is R_i , current is limited either by the maximum value supplied by the OA, or by the value $I_{max}=V_{cc}/R_L$.

---

[23] More details on this topic are given in chapter 14.1
[24] Common OA supply currents of the order of few mA. Special models can provide currents up to some A (e.g. MP38) An alternative is to use a power output buffer made by discrete components (transistor): see Appendix A.4
[25] The maximum output voltage V_o depends on the model: normally $|V_o| \approx |V_{cc}|-2V$, where $V_{cc} <30$ V . Special OA may provide larger output swing (e.g. LM143 : 130V , LME49811 :100V, MP38: 200V, MSK103 : 350V)

6.2. Floating power supply

If the OA may be powered by batteries, we may use the circuit shown in Figure 6.2a.

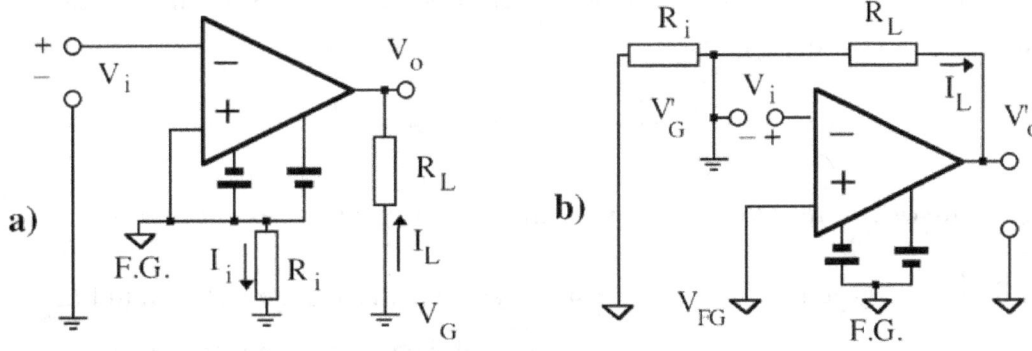

Figure 6.2

The analysis may be made easier by redrawing the circuit as shown in Figure 6.2b, where all the voltages are referred to the *floating ground* (F.G.), and we easily discover it is equivalent to a non-inverting amplifier for the signal ($-V_i$).

Applying the superposition principle to the sources V_i and $V'_G = \beta V'_o$ we get: $V'_o = -A[V_i + V'_G]$, where $V'_G = V'_o R_i / (R_i + R_L)$, or $V'_o = -A[V_i + V'_o R_i / (R_i + R_L)]$.

Rearranging the last relation we obtain $V'_o = -AV_i / [1 + AR_i / (R_i + R_L)]$, that, for $A \to \infty$, gives $V'_o = -V_i (1 + R_L / R_i)$.

The potential of the floating ground (referred to real ground) is $V_{FG} = V_i$.

Because $I_i = I_L$, $I_i = V_{FG} / R_i$ and $I_L = -V_o / R_L$ we have $I_L = -V_{FG} / R_i$.

6.3. Loaded ground with floating control voltage

If a floating control voltage is available (as a battery for d.c. signal or a transformer for a.c. signal) we may use one of the circuits shown in Figure 6.3

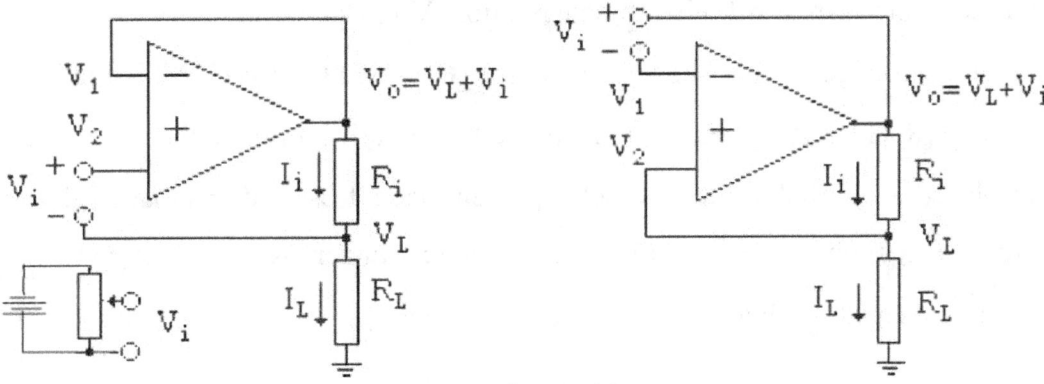

Figure 6.3

In both cases $V_o = V_L + V_i$, in the first one because $V_o = V_1$, $V_1 = V_2$ e $V_2 = V_L + V_i$, in the second one because $V_o=V_1+V_i$, $V_1=V_2$ e $V_2=V_L$. And, because $I_L=I_i$, we obtain

$$I_L=V_L/R_L=V_i/R_i.$$

Both circuits may use an OA with unipolar power supply. The maximum available current is achieved for $V_{omax} \approx V_{cc}$, that yields:

$$I_{Lmax} \approx V_{cc}/(R_L+R_i).$$

6.4. Voltage-controlled current source with all signals referred to ground

When full reference to ground is required we may use the current source shown in Figure 6.4.

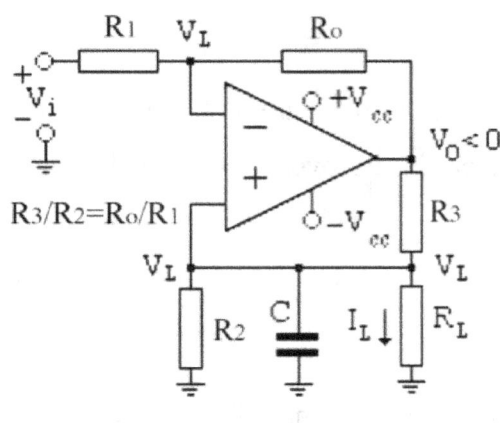

Figure 6.4

We must find the relation between V_L and V_i.

In the ideal OA model ($I_{b1}=I_{b2}=0$): the current conservation at the non-inverting node gives: $(V_o-V_L)/R_3=V_L/(R_L\|R_2)$, and at the inverting node: $(V_i-V_L)/R_1=(V_L-V_o)/R_o$.

The quantity $(V_o - V_L)$, calculated from the second relation and inserted into the first relation, and using the identity $(R_L\|R_2)=R_L R_2/(R_L+R_2)$, gives:

$$I_L = \frac{V_L}{R_L} = V_i \frac{R_2 R_o}{[R_L(R_2 R_o - R_1 R_3) - R_2 R_1 R_3]} \qquad [6.1]$$

In relation [6.1] we see that I_L still depends on the value of R_L, but if we properly choose the values of R_o, R_1, R_2, R_3, so that $R_2 R_o = R_1 R_3$ (i.e. $R_3/R_2 = R_o/R_1$) I_L is independent of R_L:

$$I_L = -V_i/R_2 \qquad [6.2]$$

The maximum available current is limited by the relation $|V_o|<|V_{cc}|$.

Considering that the voltage V_L may be written $V_L=V_o(R_L\|R_2)/(R_L\|R_2+R_3)$, from $I_L=V_L/R_L$ we get $I_{Lmax}<V_{cc}/[R_L(1+R_3/R_2)+R_3]$. To maximize I_L we must minimize R_3.

The capacitor placed in parallel to the load helps preventing self-oscillations that might be due to the positive feedback when the output is not loaded (note that for $R_L=\infty$ positive and negative feedback fraction are equal, so that the OA works at open-loop).

6.5. Voltage-controlled current source with two OA

A circuit, similar to the previous one, but with high input impedance is shown in Figure 6.5.

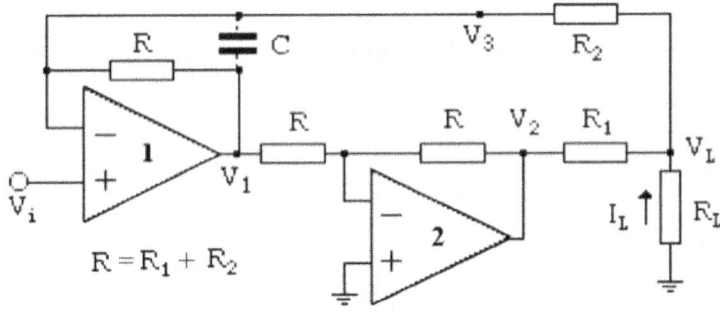

Figure 6.5

In this circuit we must carefully match the three resistors R, so that $R=R_1+R_2$,; in this situation the output current becomes independent of the load R_L : $I_L = -2V_i/R_1$.

The analysis may begin considering that OA2 is a basic inverter ($V_2 = -V_1$); OA1 acts as inverting amplifier for the source V_L with gain $(-R/R_2)$, and as non-inverting amplifier for the source V_i, with gain $(1+R/R_2)$. Moreover the voltage V_L may be calculated by superimposing the effects of the two sources V_2 and V_3: V_2 divides over $(R_1, R_L \| R_2)$ and V_3 $(=V_i)$ divides over $(R_2, R_L \| R_1)$.

The result is for V_2 and V_L:

$$V_2 = -V_1 = -[-V_L (R/R_2) + V_i(1+R/R_2)], \qquad [6.3]$$

$$V_L = V_2 \frac{R_L \| R_2}{R_1 + R_L \| R_2} + V_i \frac{R_L \| R_1}{R_2 + R_L \| R_1} \qquad [6.4]$$

By substituting [6.3] into [6.4] and dividing by R_L, we get

$$I_L = -V_i(R_2+R-R_1)/[R_L(R_2+R_1-R)+R_1R_2], \qquad [6.5]$$

Where it is clear that, for $R = R_1+R_2$, I_L becomes independent of R_L, i.e. :

$$I_L = V_L/R_L = -2V_i/R_1. \qquad [6.6]$$

The largest voltage swing takes place at the nodes at the V_1 and V_2 and must be $|V_1|=|V_2|<|V_{cc}|$: using relations [6.3] and [6.6] we obtain $I_{Lmax} < 2V_{cc}/[R_1+(R_1+R_L)(1+R_1/R_2)]$: which suggest to use small values for R_1. The capacitor is also here useful to avoid self-oscillations when load is removed. Note that here 5 resistors must be carefully matched: 3 identical R and then R_1, R_2, (with $R_1 < R_2$) such that $R_1 + R_2 = R$.

A similar circuit is shown in Figure 6.6. Here OA2 is a follower and AO1 as non-inverting summer for sources V_i and V_L scaled by the divider (R_2, R_4). Here the condition that makes I_L independent on the load R_L is $R_1/R_2 = R_3/R_4$

Because $I_5 = I_L$, we may write $I_L = (V_o - V_L)/R_5$, and we only need to calculate $V_o - V_L$.

First we note that OA1 is a non-inverting amplifier so that $V_o = (1 + R_3/R_1)V_1$. The voltage V_1 is the superposition of the sources V_i divided by (R_2, R_4), and of V_L divided by (R_4, R_2): $V_1 = V_i R_4/(R_2+R_4) + V_L R_2/(R_2+R_4)$.

The last two equations give:

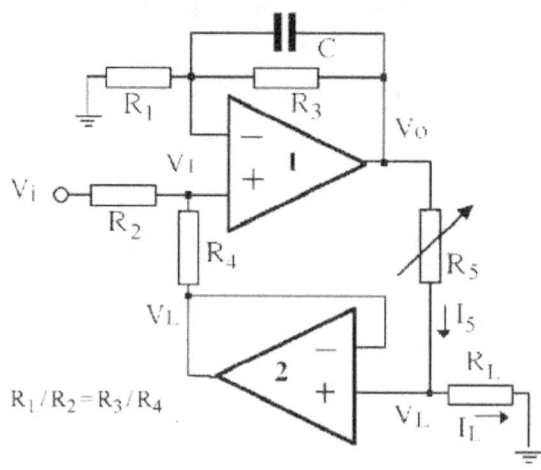

Figure 6.6

$$V_o = V_L \frac{1+R_3/R_1}{1+R_4/R_2} + V_i \frac{1+R_3/R_1}{1+R_2/R_4} \qquad [6.7]$$

If we let $R_3/R_1 = R_4/R_2$, the coefficient of V_L in [6.7] becomes 1, and observing that $(1+x)/(1+1/x) = x$, we finally get $V_o - V_L = V_i R_3/R_1$, and therefore: $I_L = V_i R_3/R_5 R_1$.

Because the value of R_5 is arbitrary, we may control the value of I_L by adjusting R_5 (using a potentiometer) instead of adjusting V_i.

The maximum achievable current I_{Lmax} is limited by the condition $|V_o| = (R_L + R_5)I_L < |V_{cc}|$, i.e. $I_{Lmax} < V_{cc}/(R_L+R_5)$. In this circuit we need to trim the value of a single resistor (e.g. R_4, once R_1, R_2 e R_3 have been chosen).

Also here the capacitor C is needed to avoid self-oscillations for $R_L = \infty$ (no load).

6.6. Current source with potentiometric control

All the previous voltage-to current converters may become current sources controlled by a potentiometer by simply using for the input voltage V_i the output voltage of a variable reference voltage source (circuits 5.1 – 5.4)

Another simpler solution is that shown in Figure 6.7, where we use unipolar power supply and load referred to ground.

The output current is $I_L = V_2/R_L$, the same flowing across R ($I_b=0$). Because the OA is a follower $V_2=V_1=V_o$, and the voltage drop across R is the same as that across the zener V_z. As a consequence the output current is $I_L=V_z/R$, where R may be freely adjusted (above a minimum value R_{min}, as well see). The maximum current is $I_{Lmax}=(V_{cc}-V_z)/(R_L+R_o)$, so that a good choice is $V_z \ll V_{cc}$, while R_o should be

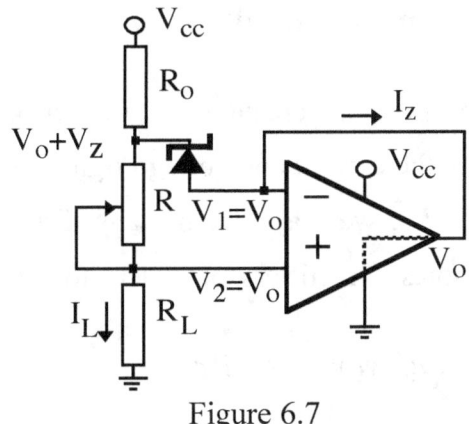

Figure 6.7

selected to provide $I_{zmin} < I_z < I_{AOmax}$, where I_{zmin} is the threshold zener current, and I_{AOmax} is the maximum current available at the OA output.

Because $I_z = (V_{cc}-V_z)/R_o - I_L$ the limits above defined for I_z set limit values to R_o as follows:

$(V_{cc}-V_z)/(I_{AOmax}+I_{zmin}) < R_o < (V_{cc}-V_z)/I_{zmin}$.

The minimum value for R which corresponds to I_{Lmax}, is set by the condition:

$$V_z/R_{min} \approx (V_{cc}-V_z)/(R_L+R_o), \quad \text{i.e.} \quad R_{min} > (R_L+R_o)V_z/(V_{cc}-V_z).$$

7. Non linear circuits

In the previous chapters were described several circuits essentially made of AO and resistors, where the current and voltage signals are processed linearly. By introducing *non-linear elements*, as *diodes*, we may obtain many different non-linear devices. In this chapter we analyze some examples of rectifiers, peak detectors, and basic logarithmic and exponential amplifiers.

7.1. Half-wave rectifier

The rectifier is a device that passes positive signals and blocks negative signals. The diode, by itself is a basic half-wave rectifier, because it approximates an *unipolar switch*, i.e. a switch driven by the sign of its bias[26].

The transfer function f of an *ideal half-wave rectifier* $V_o = f(V_i)$ is $f(V_i) = V_i$ for $V_i > 0$ and $f(V_i) = 0$ for $V_i < 0$.

In Figure 7.1a. we show the passive circuit that approximates an *half-wave rectifier*.

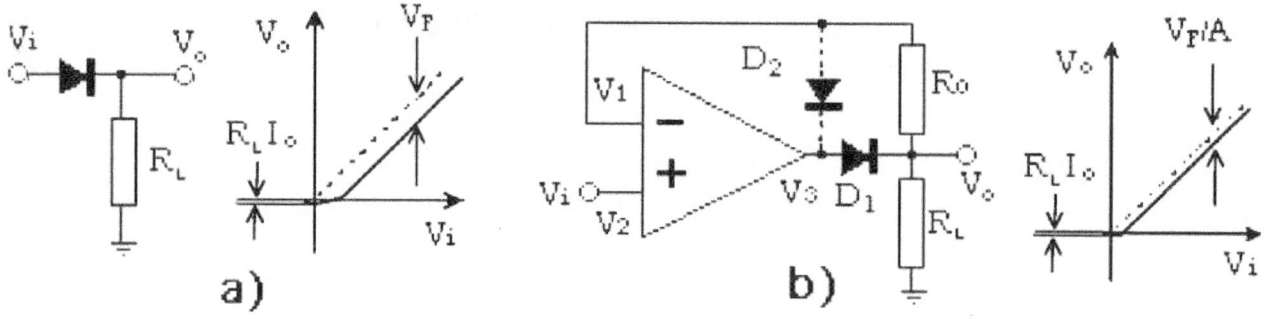

Figure 7.1

Figure 7.1a shows that the transfer function of this circuit differs from the one of the ideal half-wave rectifier. The negative signal is cut almost completely: a small part is left $V_o = -R_L I_o$, where I_o is the *diode reverse* current (of the order of few µA). For positive input voltages the output only approximates the ideal function $V_o = V_i$, the effective output is $V_o = V_i - V_F < V_i$, where $V_F \approx 0.6V$ is the voltage drop across the *forward biased diode*.

The circuit of Figure 7.1b is a better approximation of an ideal rectifier: by inserting the diode D1 into the OA feedback, we strongly reduce the effect of the forward voltage drop V_F.

To understand it we first neglect the diode D2. When $V_i > 0$, the output V_3 tends to rise up to the voltage AV_i (where A is the open-loop gain). But the diode D1, forward-biased, feeds the positive voltage to the inverting input through R_o. The negative feedback blocks V_2 at the value

---

[26] For more details on the diode see appendix A.1.

$V_2 = V_i$, so that $V_3 = V_i + V_F$. Because $I_b \approx 0$ the voltage drop across R_o is negligible and $V_o \approx V_i$.

The effective error may be calculated by taking into account the real *finite* value of the open-loop gain A. The differential input voltage is $\varepsilon = V_1 - V_i = V_F/A$. By considering also the finite bias current: $V_o = V_i - (R_o I_b + V_F/A) = V_i - \Delta V$. For example with $A \approx 10^{+5}$, $I_b \approx 0.1\,\mu A$, $R_o \approx 1\,k\Omega$ we have $\Delta V \approx 0.1\,mV$.

For $V_i < 0$, the negative feedback is lost and the output voltage V_3 of the OA saturates: $V_3 = -V_{cc}$. The diode D1 is switched off and the reverse diode current I_o gives: $V_o = -R_L I_o$.

To avoid possible *latch-up*, i.e. the freezing of the OA at saturation (see chapter 9), we may establish negative feedback through the diode D2, that for $V_i < 0$, gives $V_2 = V_i$ and $V_3 \approx V_i - V_F$. In this case, however, for $V_i > 0$, we get $V_o = V_i R_L/(R_o + R_L)$, and we must use a very high value for the resistor R_o ($R_o \gg R_L$) to approximate the output to zero: $V_o = V_i R_L/(R_o + R_L)$.

This choice requires using a FET-input OA to retain the ideal OA approximation (I_b negligible with respect to the feedback current $I_r = V_i/R_o$).

This problem is avoided by the *inverting* half-wave rectifier shown in Figure 7.2.

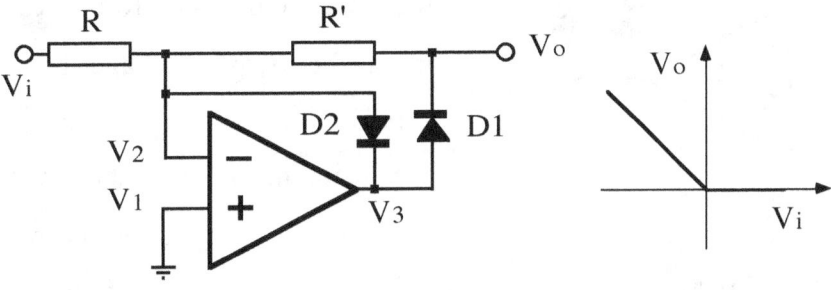

Figure 7.2

Here, for $V_i < 0$ ($V_3 > 0$) the negative feedback is supplied by R' and D1, while D2 is reverse-biased and can be neglected. In the ideal OA approximation ($I_b = 0$, $A = \infty$) we have $V_1 = V_2 = 0$, $V_o/R' = -V_i/R$, i.e. $V_o = -(R'/R)V_i$ and $V_3 = V_o + V_F$. With $R = R'$ we get $V_o = -V_i$.

For $V_i > 0$ the negative feedback is supplied by D2 (D1 reverse biased): $V_1 = V_2 = 0$, $V_3 = -V_F$, and $V_o = V_2 - R'I_o \approx 0$.

7.2. Full-wave rectifier

A full-wave rectifier has the transfer function $V_o = |V_i|$. One example is shown in the circuits of Figure 7.3 with two OA: the first one has a twin-diode feedback and the second one is a basic differential amplifier. Diode D1 switches-on for positive input and diode D2 for negative input.

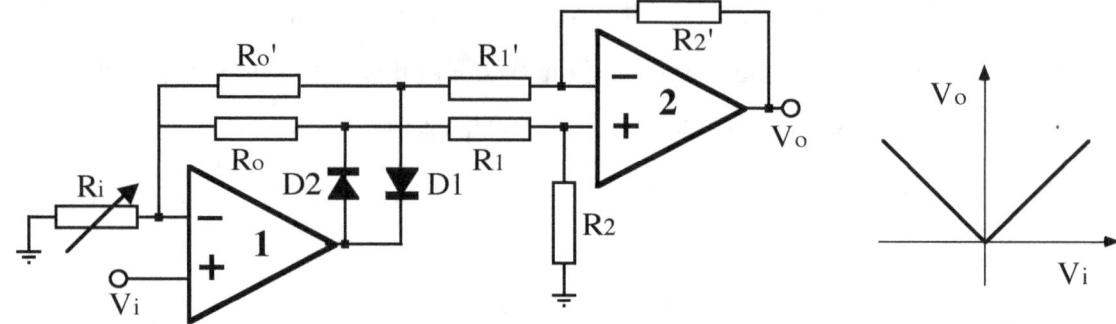

Figure 7.3

Two conditions must be satisfied: $R_o'=R_o$, $R_1/R_1'=R_2/R_2'$ to give $V_o=(R_oR_2/R_iR_1)|V_i|$. The gain $G=(R_oR_2/R_iR_1)$ may be set by a single resistor (R_i), so we may also get $G=1$.

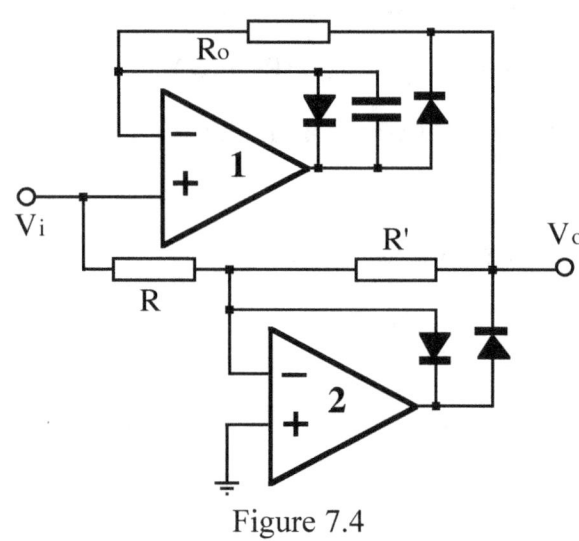

Figure 7.4

Another full-wave rectifier that requires matching only two resistors ($R=R'$) is shown in Figure 7.4: it is made of the circuits of Figure 7.1b and 7.2, placed in parallel.

OA1 is a follower for positive input and OA2 an inverter for negative input. Here we may release the condition $R_o \gg R_L$, because for negative input the output voltage is set by OA2. The capacitor helps rejecting self-oscillation increasing negative feedback when $V_i > 0$.

A third full-wave rectifier is shown in Figure 7.5. Here OA1 is the inverting half-wave rectifier described in Figure 7.2, that gives $V_1 = -(R'_1/R_1)V_i$, for $V_i > 0$, and $V_1 = 0$ for $V_i < 0$. The voltage V_1 is added to the input V_i by the inverting summer AO2. If the resistors satisfy the condition $R_2 = (R_i/2)(R'_1/R_1)$ the output voltage is $V_o = (R_o/R_i)|V_i| = G|V_i|$.

A simple choice is $R'_1 = R_1 = R_i = R$ and $R_2 = R/2$, that, for $R_o = R$ gives $G=1$.

The circuit of Figure 7.5, however, does not offer high input impedance. An alternative full-wave rectifier with high input impedance is shown in Figure 7.6.

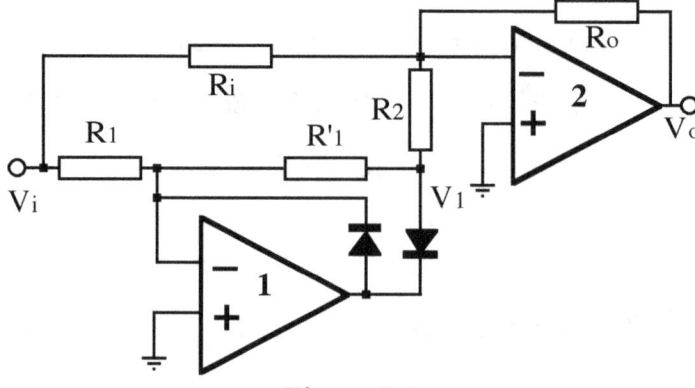

Figure 7.5

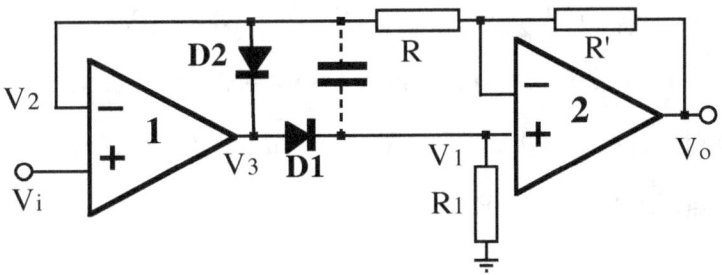

Figure 7.6

Here OA2 works as inverter for the signal $V_2=V_i$, with $G=-R'/R$, when $V_i<0$ because the D2 feed full negative feedback to OA1 (follower). When $V_i>0$, we get $V_i=V_2=V_1$ and no current flows across R and R', so that $V_o=V_i$. If $R=R'$ we get $V_o=|V_i|$.

Resistor R_1 is needed to forward bias the diode D1 for $V_i>0$, and to supply the bias current I_{b+} to AO2 when $V_i<0$.

A simple variant of the previous circuit is shown in Figure 7.7, where the resistor R_2 is added at the inverting input of OA1, imposing however a gain $G>1$ for the full-wave rectifier.

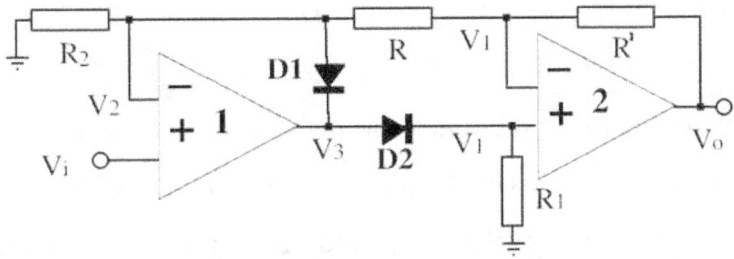

Figure 7.7

For $V_i<0$ we find again $V_o=-(R'/R)V_i=-GV_i$.

For $V_i>0$, because diode D1 is reverse biased, by neglecting its reverse current we may write the current conservation along the resistors R, R' and R_2: $V_i/R_2=(V_1-V_i)/R=(V_o-V_1)/R'$, where we also used the equation $V_2=V_i$. By eliminating the variable V_1 we obtain: $V_o=[1+(R+R')/R_2]V_i=[1+(G+1)R/R_2]V_i$, where $G=R'/R$. By choosing $R_2=R(G+1)/(G-1)$ we obtain $V_o=G|V_i|$. Here always $G>1$, because for $G=1$ must be $R_2=\infty$ i.e. again circuit 7.6.

Another example of full-wave rectifier with high input impedance is shown in Figure 7.8.

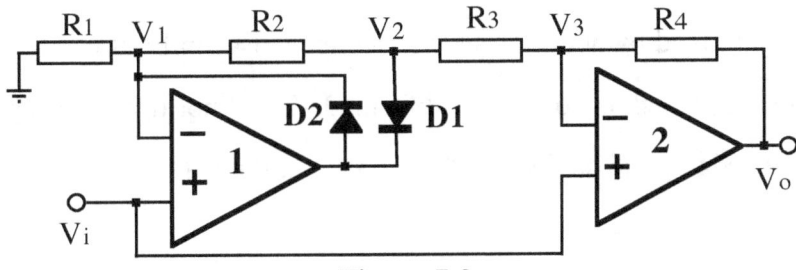

Figure 7.8

For $V_i<0$ the diode D2 is switched-off and the non-inverting OA1 gives: $V_2=(1+R_2/R_1)V_i$, while OA2 amplifies both V_i and V_2: $V_o = (1+R_4/R_3)V_i - (R_4/R_3)V_2 = (1 - R_4R_2/R_3R_1)V_i$.

For $V_i > 0$, the negative feedback establish $V_1 = V_2 = V_3 = V_i$, and no current flows across resistors R_2, R_3 (and R_4), so that $V_o = V_i$. In order to achieve $V_o = |V_i|$ we must set $R_4R_2/R_3R_1 = 2$, e.g. $R_4/2 = R_2 = R_3 = R_1$.

7.3. Peak detector

An half-wave rectifier loaded by a capacitor becomes a *peak detector* for positive input voltages. An example is given in Figure 7.9a, and Figure 7.9b shows the time evolution of input (dashed line) and output (full line) voltages.

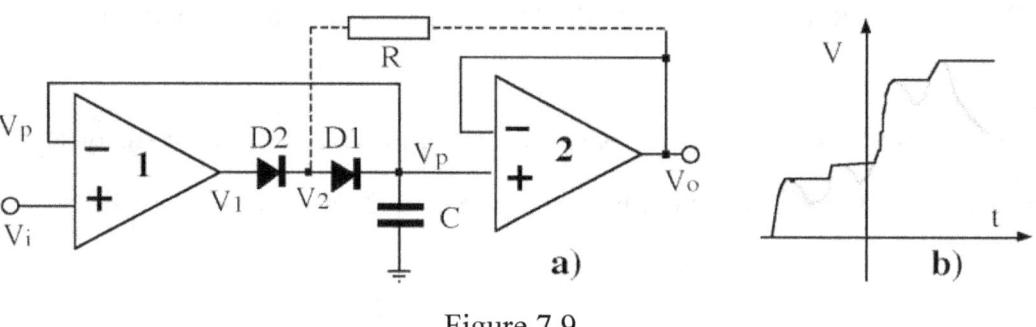

Figure 7.9

In the analysis of this circuit we first neglect the resistor R, and we assume only positive input voltages. Within the ideal OA model ($I_b = 0$), the capacitor is charged through the diodes D2 and D1 to the peak value V_p of the input voltage V_i, and the output V_o keeps this value also when V_i becomes smaller than V_p, assuming $I_o = 0$ for the diodes reverse current. With real diodes ($I_o \neq 0$) when $V_i < V_p$, the voltage V_1 saturates at $-V_{cc}$, and the capacitor C slowly discharges through the diodes reverse biased. Adding the resistor R, the voltage V_2 is held to the peak value $V_p = V_o$, by the negative feedback of OA2.

There is no more voltage drop across D1 and the reverse current vanishes, so that the capacitor holds its charge (if we still neglect the bias input current of OA1). The reverse current of D2 is supplied by OA2 through R.

A negative peak detector is obtained by reversing the two diodes: the output voltage keeps the minimum values assumed by negative input.

This circuit may be improved by adding a second feedback (R_2 in Figure 7.10) which cancels the effect of finite I_b in OA1, which is now drained from OA2 and not from the capacitor C.

The third diode D3 speeds-up the device by blocking V_1 at the value $V_p - V_F$.

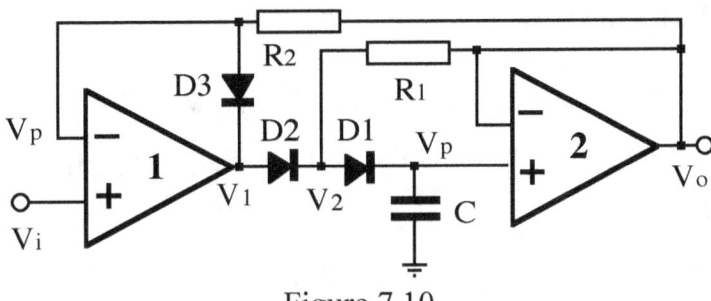

Figure 7.10

OA1 should be selected for high differential input and OA2 for low bias currents.

7.4. Logarithmic and exponential amplifiers

Logarithmic and exponential amplifiers allow multiplication and division of analogic signals, and they could be used to build analogic computers. Their more common application, however is for signal compressing or expanding, in order to change the reading scale. For analog multiplication and division the most used devices are the *transconductance* IC[27].

Her we give only a brief analysis of the working principle of logarithmic and exponential amplifiers in basic examples.

To understand the behavior of the following circuits we must refine the approximation of the diode used until now (the *unipolar switch model*), adopting the *ideal diode model*[28]. The ideal is as a non-linear element defined by the following voltage/current relation:

$$I_d(V) = I_o \exp(qV/K_BT) \qquad [7.1]$$

where V is the *forward voltage*, I_d the *forward current*, and I_o the *reverse current* (or *leakage current*); $K_B = 1.38 \times 10^{-23}$ J/K is the Boltzmann constant, T the temperature in Kelvin, $q = 1.6 \times 10^{-19}$ Coulomb the electronic elementary charge. At room temperature (≈ 300 K) $K_BT/q \approx 26$ mV. This approximation is good until $V \gg K_BT/q$, i.e. $I_d \gg I_o$.

7.4.1. Logarithmic amplifier

By replacing the feedback resistor with a diode in an inverting amplifier, as in Figure 7.11, we obtain an output voltage proportional to the logarithm of the input voltage Vi, assuming Vi>0.

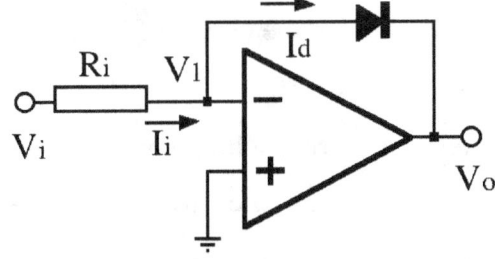

Figure 7.11

---

[27] Tranconductance multipliers and dividers are treated in detaild i in *Linear Integrated Circuit Applications*, G.B. Clayton, chapt. 6, in *Operatinal amplifier and applications*, W.G. Young chapt 6, and in *Introduction to Operational Amplifiers: Theory and Applications*, J. Wait et al., chapt 3.

[28] For more details on the ideal diode model see Appendix A.1.

In fact, by neglecting I_b, for the ideal diode we obtain:

$$I_i = V_i/R_i = I_d = I_o \exp(qV/K_BT) \quad [7.2]$$

From Figure 7.11 we have $V_d = V_1 - V_o$, and therefore $V_i/R_i I_o = \exp(-qV_o/K_BT)$.

Taking the logarithm:

$$V_o = -(K_BT/q) \ln(V_i/I_oR_i) = -2.3(K_BT/q)\log_{10}(I_i/I_o) = -S(\log_{10} I_i - \log_{10} I_o) \quad [7.3]$$

where the scale factor $S = 2.3(K_BT/q)$ depends on temperature with slope $\partial S/S \partial T = 0.003 \, °C^{-1}$.

The temperature dependence is also contained in the term $\log_{10} I_o$ which approximately duplicates every 10 °C; moreover the magnitude of I_o depends on the diode type, ranging from 1 nA to 1 µA.

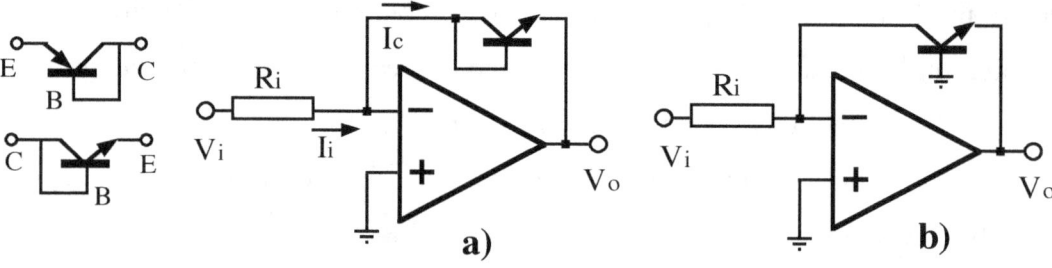

Figure 7.12

The ideal $V_d (I_d)$ curve is normally obeyed by real diodes for max three decades in I_d. For an extended range we may use a transistor *connected as a diode*, i.e. with the collector shorted to the base electrode, as in Figure 7.12a.

Another configuration, also named *transdiode*[29], is shown in Figure 7.12b. Here the collector and base electrodes of the transistor are kept at the same voltage through the negative feedback (collector at virtual-ground) so that the effective behavior is the same of ideal diode.

In the circuit of Figure 7.12b the I_d covers up to 10 decades (up to a few mA) and the output voltage spans about 0.6 V.

The scale factor ($S \approx 60$ mV/decade) may be changed using the circuit of Figure 7.13.

We have $V_d = -V_1$, and neglecting the feedback current with respect to the current flowing across the divider (R_1, R_2), we obtain $V_1 = V_o R_1/(R_1 + R_2)$, i.e.:

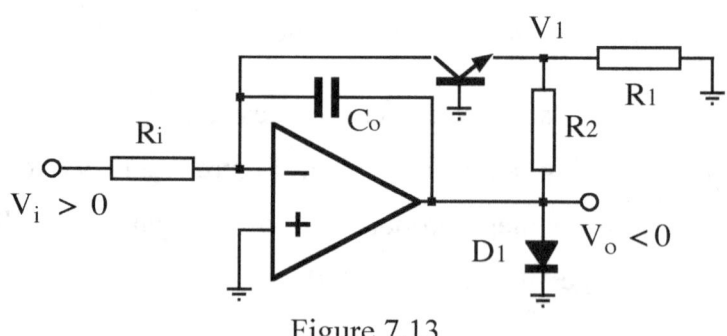

Figure 7.13

---

[29] For more details on transistors see Appendix A.3 and for transdiodes see *Operational Amplifiers*, G.B. Clayton, chapt. 5.

$$V_0 = -\frac{R_1+R_2}{R_1} 2.3\left(\frac{K_B T}{q}\right) \log_{10}(I_i/I_0) \approx -\left(1+\frac{R_2}{R_1}\right) 60 \log_{10}(V_i/R_1 I_0) \quad [\text{mV/decade}].$$

The scale factor S, for $R_2 \approx 16\, R_1$, becomes $S \approx 1$ V/decade. Moreover, using for R_1 a PTC thermistor with temperature coefficient $\approx 0.3\%$ °C^{-1}, S becomes temperature independent.

In Figure 7.13 the capacitor C_o helps avoiding self-oscillations and the diode D1 protects the transistor that could burn under excessive reverse bias. A more complete analysis should account for the bias currents I_b of the OA. [30]

By assembling two logarithmic amplifiers and one differential amplifier we obtain an *analog divider* (figure 7.14).

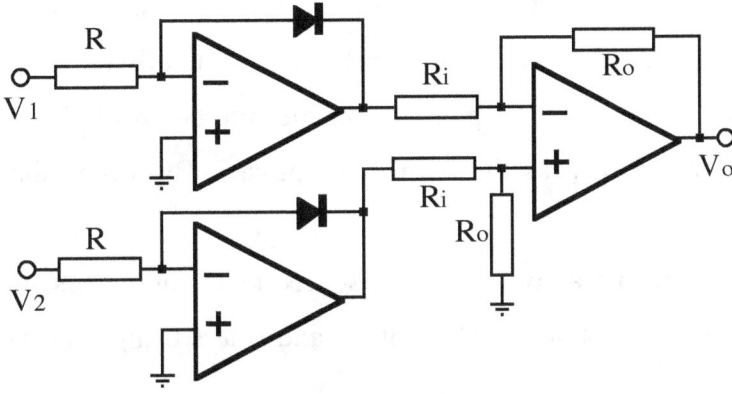

Figure 7.14

The output voltage is: $V_0 = \frac{R_0}{R_1}\left(\frac{K_B T}{q}\right) \ln(V_2/V_1)$. Note that here the dependence of the diode leakage current I_o vanishes.

7.4.2. Exponential amplifier

An exponential amplifier can be obtained from the circuit of Figure 7.11 by interchanging diode and resistor in the feedback network : the result is shown in Figure 7.15.

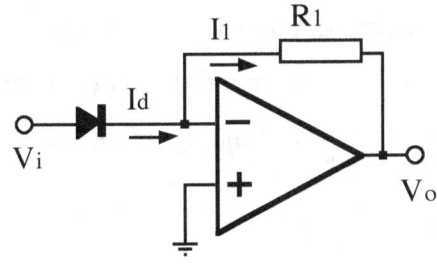

Figure 7.15

Using the ideal OA model, and for input voltages satisfying the relations $0.1 < V_i < 0.6$ V and $I_1 = I_d < 1$ mA we get the output voltage: $V_o = -R_1 I_1 = -R_1 I_o \exp(qV_i/K_B T)$.

---

[30] More details on logarithmic amplifiers are given in Operational Amplifiers, G.B. Clayton, chapt. 5, or *Operational Amplifiers and Applications*, W.G. Young, chapt. 6.

8. Active filters

In this chapter we analyze filters, i.e. circuits whose transfer function *depends on the frequency*. The *transfer function* is the ratio between the output signal and the input signal A filter modifies both the *amplitude* and the *phase* of sinusoidal signals: in mathematical language, we may say that the transfer function of a filter is a *complex* function[31]. In a low-pass filter, for example, the low frequency signals remain unchanged while high frequency signals are attenuated.

Examples of passive filters are the capacitive and inductive dividers (RC and RL filters), described in Appendix B. The active filters offer the advantages of low output impedance and high input impedance, and they may also have gain G>1.

In the literature we may find many recipes for designing filters with any transfer function (Butterworth filters, Tchebeyscheff filters, Bessel filters). Generally the filters are classified by an *order number n* (with $n=1,2,3,4...$) depending on the number n of the *poles* of their transfer function[32]; where n may be seen as the number of passive RC filters that should cascaded to approximate such filter.

In this chapter we analyze the active filters most frequently used: first order filters, multiple-feedback filters, VCVS filters, state-variable filters, and filters using impedance converters (NIC, gyrators).

The first order filters are the low-pass (integrator) and the high-pass (differentiator); the all-pass (phase-shifter) will be described in §10.3.

The multiple-feedback filters, and VCVS filters (*Voltage Controlled Voltage Source*) here described will be those of order 2: higher order filters are generally obtained by cascading filters of this type. The state-variable filters use the technique of analog calculators and are made of active integrators and summers. The circuits NIC (*Negative-Immittance-Converter*) use OA with both positive and negative feedback to transform an impedance (Z) into its negative (–Z), and gyrators convert (Z) into its reciprocal (1/Z).

8.1. Active Integrator

By replacing the feedback resistor R_o in the inverting amplifier of Figure 3.1 with a capacitor we obtain an active integrator.

---

[31] Complex (or vectorial) notation of signals is briefly treated in Appendix B. See also http://en.wikipedia.org/wiki/Complex_number A time-dependent voltage signal V(t) may always be seen as superposition of a large (or infinite) number of sinusoidal signals, and it may therefore be represented by a function which is a sum of sine waves. In the *complex* notation the sinusoidal signal is Vexp(jωt) = V(cos ωt + j sin ωt), where j is the imaginary unit.

[32] A definition of poles and zeroes in a transfer function is given in Appendix B.4

In Figure 8.1a, the voltage $V_c(t)$ across the capacitor C changes with time t due to the charge q(t) carried by the current $I_c(t)$.

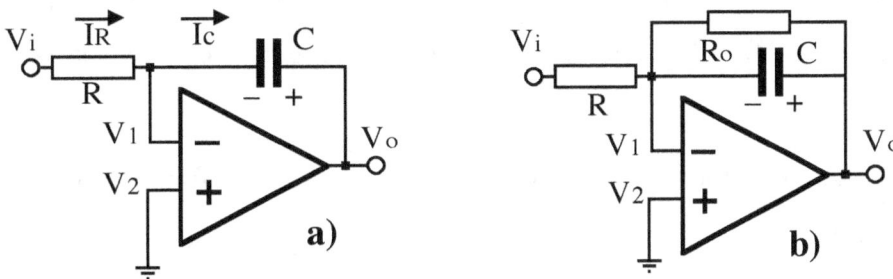

Figure 8.1

In the ideal OA model ($I_b=0$) we get $I_c(t)=I_R(t)=V_i(t)/R$.

Because $V_c(t) = -[V_1 - V_o(t)]$, from $V_1 = V_2 = 0$, we obtain:

$$V_o(t) = -V_C = \frac{-q(t)}{C} = -\frac{1}{C}\int I_C(t)dt = -\frac{1}{RC}\int_0^t V_i(t)\,dt + V(0)$$

where we used the definition $I(t) = \partial q(t)/\partial t$. The product $\tau = RC$, named *time constant*, is the time required to bring the output voltage from zero to the same constant voltage applied to the input.

In Figure 8.1b the resistor R_o, in parallel with C, provides the necessary d.c. feedback: without R_o the finite input bias current (and input offset voltage) of real OA produce an output that brings the output (even with $V_i = 0$) at saturation (positive or negative depending on the V_{os} and I_b values).

For a.c. signals it is better to describe the circuit response through the transfer function $T(j\omega) = V_o/V_i$. The capacitor impedance[33] is ($Z_c = 1/j\omega C$), so that the ideal integrator of Figure 8.1a may be seen as an inverting amplifier with complex feedback with $G(j\omega) = V_o/V_i = -Z_C/Z_R$:

$$T(j\omega) = -1/j\omega RC,$$

Therefore the OA saturates at zero frequency, i.e. the output drifts to $\pm V_{cc}$ for any d.c. input voltage, making this circuit useless for any practical application.

By introducing the resistor R_o the transfer function becomes:

$$T(j\omega) = -\frac{Z_C \| Z_{R_o}}{Z_R} = -\left(\frac{R_0}{R}\right)\frac{1/(R_0 C)}{1/(R_0 C) + j\omega} = G\frac{\omega_0}{\omega_0 + j\omega}. \qquad [8.1]$$

where $\omega_0 = 1/R_o C$ is named *cut frequency*.

The module of $T(j\omega)$ is $|T(\omega)| = \omega_0 G / \sqrt{\omega^2 + \omega_0^2}$, and the phase is $\phi = \arctan(-\omega/\omega_0)$. From now-on

---

[33] The complex impedance id described in more details in Appendix B.

the module of the transfer function will be named $A(\omega) = |T(s)|$. For d.c. signals or low-frequency signals ($\omega \ll \omega_0$), $A(\omega)$ is practically constant $A(\omega) \approx G = -(R_o/R)$, while for high frequencies ($\omega \gg \omega_0$) the transfer function approximates the one of the previous circuit: $T(j\omega) \approx -1/j\omega RC$. The phase shift at high frequency is $-\pi/2$. At the cut frequency ω_0 we get: $A(\omega) = G/\sqrt{2}$ and $\phi(\omega_0) = -\pi/4$.

The integrator is therefore a low-pass filter of order 1 (the transfer function has one pole, i.e. one zero at the denominator).

8.2. Differentiator

By replacing the input resistor R_i in the inverting amplifier of Figure 3.1 with a capacitor we obtain the active differentiator of Figure 8.2a.

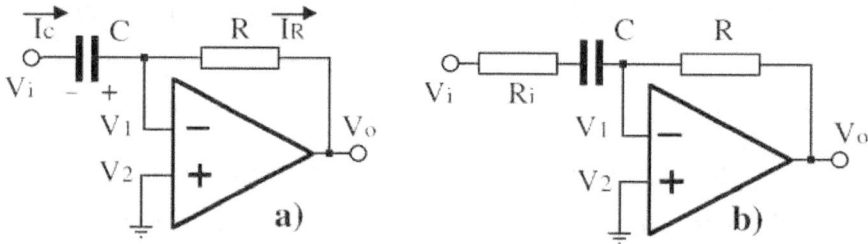

Figure 8.2

Because $I_b = 0$, the capacitor C is charged by the current $I_c(t) = I_R(t) = \partial q(t)/\partial t$, where $q(t)$ is the charge accumulated on the capacitor electrodes, and the voltage $V_c(t) = [V_i(t) - V_1] = q(t)/C$. The ideal OA model ($V_1 = V_2 = 0$), gives $V_o(t) = -RI_c(t) = -R \, \partial q(t)/\partial t$, and therefore:

$$V_o(t) = -RC \, \partial V_i(t)/\partial t$$

The transfer function is $T(j\omega) = -R/Z_c(j\omega) = -j\omega RC$,: $A(\omega) = |T(j\omega)|$ is zero for $\omega = 0$ and increases linearly with frequency. This enhances the high frequency noise, making this circuit not practically usable. A substantial improvement is obtained by adding an input resistor R_i as in Figure 8.2b The new transfer function becomes (by simplifying the notation with $s = j\omega$) :

$$T(s) = -\frac{R}{Z_c(s) + R_i} = -\left(\frac{R}{R_i}\right)\left(\frac{sR_iC}{1+sR_iC}\right) = \left(-\frac{R}{R_i}\right)\frac{s}{1/R_iC + s} = G\frac{s}{\omega_0 + s} \qquad [8.2]$$

Here the cut frequency is $\omega_0 = 1/R_iC$, and the gain, still increasing with frequency, saturates at $G = -R/R_i$ at frequencies $\omega \gg \omega_0$. More precisely, we get $A(\omega) = |T(s)| = \omega G/\sqrt{\omega^2 + \omega_0^2}$, and $\phi = \arctan(\omega_0/\omega)$, i.e. the phase shift becomes $+\pi/2$ for $\omega \gg \omega_0$. At the cut frequency $A(\omega_0) = G/\sqrt{2}$, and $\phi(\omega_0) = +\pi/4$. The differentiator is a high-pass filter of order 1.

8.3. Multiple feedback filters

Multiple feedback filters of second order are made by one OA ad a passive network with impedances Z_i (R and C) in the general layout of Figure 8.3.

The transfer function of this circuit may be easily calculated by observing that node B is a virtual ground ($V_B=0$), $I_3=I_5$ because $I_b=0$, and by imposing the current conservation at node A : $I_1=I_2+I_3+I_4$, and at node B: $I_3=I_5$.

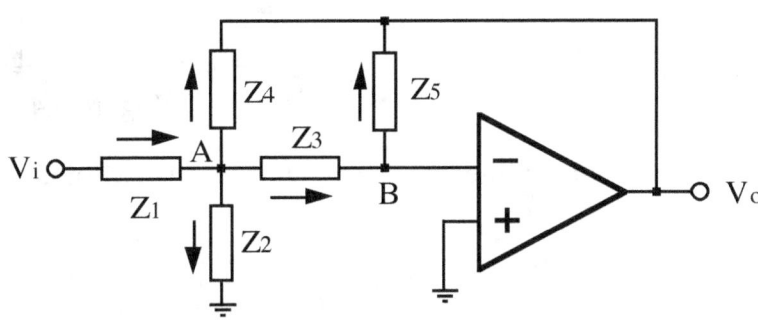

Figure 8.3

The first equation (node A) may be written, using Ohm's Law:

$$I_1 = -\frac{V_i - V_A}{Z_1} = \frac{V_A}{Z_2} + \frac{V_A}{Z_3} + \frac{V_A - V_o}{Z_4} = I_2 + I_3 + I_4$$

At node B we have:

$$I_3 = \frac{V_A}{Z_2} = -\frac{V_o}{Z_5} = I_5,$$

which gives $V_A = -V_o Z_3/Z_5$; replacing V_A in the first equation and solving for V_o we get $T(s) = V_o/V_i$:

$$T(s) = -\frac{Z_4/Z_1}{(Z_3 Z_4)/(Z_2 Z_5) + (Z_3 + Z_4 + Z_3 Z_4/Z_1)/Z_5 + 1} \quad [8.3]$$

where we wrote the complex impedances $Z_i(s)$ simply as Z_i

This is the *general form* of T(s) for all the *second order multiple feedback filters*, that we'll use to obtain the particular T(s) in special cases.

We will analyze the three main cases: low-pass filter, high-pass filter and band-pass filter.

8.3.1. Low-pass filter

If in the circuit of Figure 8.3 Z_1, Z_3, Z_4 are resistors (Z=R), and Z_2, Z_5 are capacitors (Z=1/sC), we obtain a low-pass filter (figure 8.4) with transfer function :

$$T(s) = -\frac{R_4/R_1}{s^2 R_3 R_4 C_2 C_5 + s C_5 (R_3 + R_4 + R_3 R_4/R_1) + 1} = \frac{-G\omega_0^2}{s^2 + 2s\zeta\omega_0 + \omega_0^2} \quad [8.4]$$

where $G = R_4/R_1$, $\omega_0 = 1/\sqrt{R_3 R_4 C_2 C_5}$ and $\zeta = \omega_0 C_5 (R_3 + R_4 + R_3 R_4/R_1)/2$ is named *damping factor*.

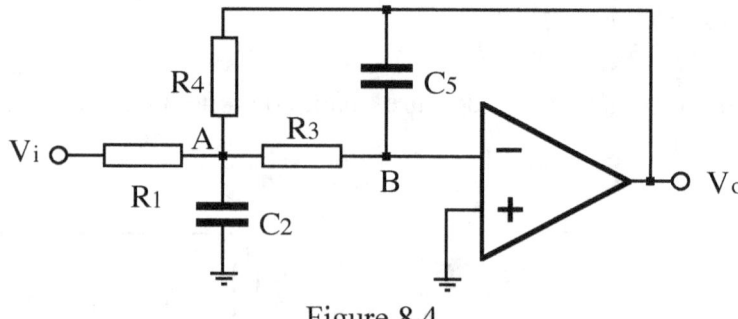

Figure 8.4

In this low-pass filter the frequency dependence of amplitude and phase are:

$$A(\omega) = \frac{G\omega_0^2}{\sqrt{(\omega_0^2-\omega^2)^2+(2\zeta\omega\omega_0)^2}} \quad , \quad \varphi(\omega) = \text{arctg}\frac{-2\zeta\omega\omega_0}{\omega^2-\omega_0^2} \;.$$

8.3.2. The high-pass filter

If in the circuit of Figure 8.3, Z_1, Z_3, Z_4 are capacitors ($Z=1/sC$) and Z_2, Z_5 are resistors ($Z=R$), we obtain a low-pass filter (figure 8.5) with transfer function:

$$T(s) = \frac{-s^2 C_1/C_4}{1/(C_3C_4R_2R_5)+s(1/C_3+1/C_4+C_1/C_3C_4)/R_5+1} = \frac{-s^2 G}{s^2+2s\zeta\omega_0+\omega_0^2}, \quad [8.5]$$

where $G=C_1/C_4$, $\omega_0=1/\sqrt{R_2R_5C_3C_4}$, and $\zeta=(1/C_3+1/C_4+C_1/C_3C_4)/(2R_5\omega_0)$.

$$A(\omega) = \frac{G\omega^2}{\sqrt{(\omega_0^2-\omega^2)^2+(2\zeta\omega\omega_0)^2}} \quad , \quad \varphi(\omega) = \text{arctg}\frac{2\zeta\omega\omega_0}{\omega^2-\omega_0^2} \;.$$

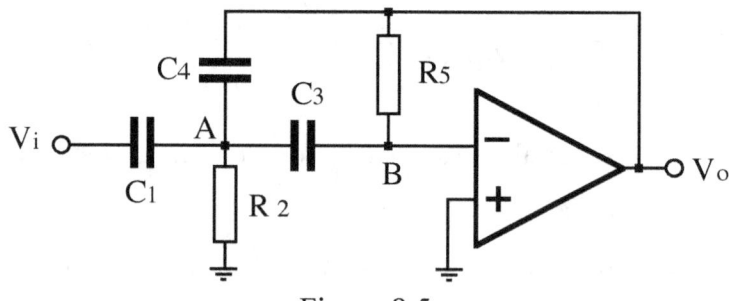

Figure 8.5

8.3.3. The band-pass filter

If in the circuit of Figure 8.3, Z_1, Z_2, Z_5 are resistors and Z_3, Z_4 are capacitors, we obtain a band-pass filter (figure 8.6) with transfer function:

$$T(s) = \frac{-s^2/(R_1C_4)}{1/(C_3C_4R^*R_5)+s/(C^*R_5)+s^2} = \frac{-sG\omega_0/Q}{s^2+s\omega_0/Q+\omega_0^2} \quad [8.6]$$

where $C^*=(C_3C_4)/(C_3+C_4)$ and $R^*=R_1\|R_2$, $\omega_0=1/\sqrt{R^*R_5C_3C_4}$, $Q=\omega_0C^*R_5$ is the *quality* *factor*[34], and $G=(R_5C^*)/(R_1C_4)$ is the gain.

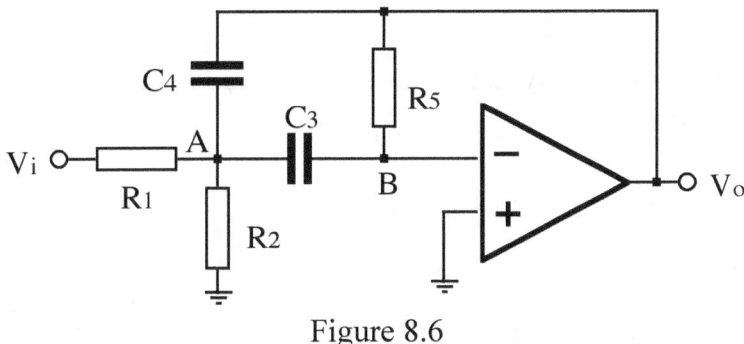

Figure 8.6

The amplitude is $A(\omega)=|T(s)|=G/\sqrt{1+Q^2(\omega/\omega_0-\omega_0/\omega)^2}$, and the phase shift, which change sign at $\omega=\omega_0$, is $\phi(\omega)=\arctan[-Q(\omega/\omega_0-\omega_0/\omega)]$.

8.4. Quality factor and damping factor

The meaning of the damping factor ζ is explained by the graphs of Figure 8.7 where the amplitude $A(\omega) = |T(s)|$ (normalized to G) is plotted vs. frequency (normalized to the frequency ω_0) for various values of ζ (for high-pass and low-pass). For small ζ values the filter response is peaked near the frequency ω_0. The peak frequency ω_p may be obtained by zeroing the first derivative of $A(\omega)$: $\omega_p=\omega_0\sqrt{1-2\zeta^2}$ for the low-pass, and $\omega_p=\omega_0/\sqrt{1-2\zeta^2}$ for the high-pass. This shows that a peak appears only for $\zeta<1/\sqrt{2}\approx 0.7$.

The peak-amplitude is $A(\omega_p)=G/\left(2\zeta\sqrt{1-2\zeta^2}\right)$.

The peak disappears in the *Butterworth type* filter where $\zeta=1/\sqrt{2}$. At the cut-frequency ω_0, we get

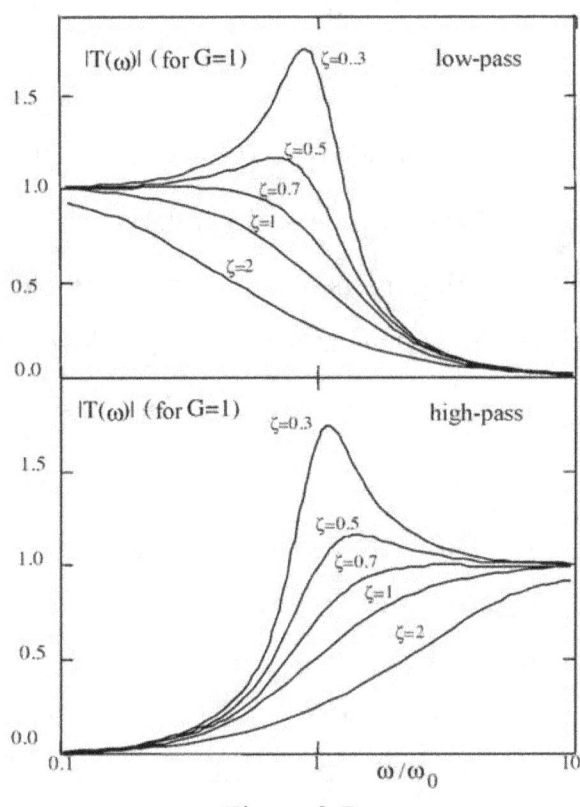

Figure 8.7

$A(\omega_0)=G/2\zeta$; in the Butterworth filter therefore $A(\omega_0)=G/\sqrt{2}$.

The band-pass filter is better described by the parameter $Q=(2\zeta)^{-1}$.

---

[34] For the quality-factor see also: http://en.wikipedia.org/wiki/Q_factor

Figure 8.8 gives the band-pass response for different Q values as a function of ω/ω_0.

Note that the transfer function is symmetric with respect to ω_0 if the abscissa is traced in log-scale. In the band-pass filter ω_0, takes the name of *central frequency*, and we find that $A(\omega_0) = G$.

The larger is Q, the narrower is the peak in the band-pass response: the quality factor Q is defined as $Q = \omega_0/(\omega_2 - \omega_1)$, where ω_1 and ω_2 are the frequencies at which $A(\omega_{1,2}) = A(\omega_0)/\sqrt{2}$.

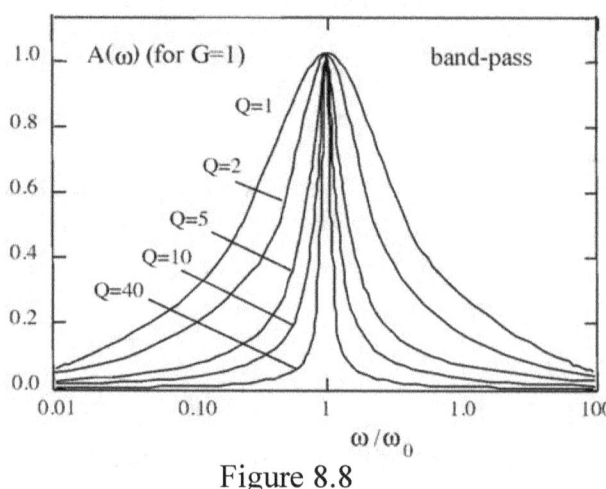

Figure 8.8

In fact the equation $A(\omega) = G/\sqrt{1 + Q^2(\omega/\omega_0 - \omega_0/\omega)^2}$, letting $A(\omega_{1,2}) = A(\omega_0)/\sqrt{2}$, becomes $1 + (\omega_0/\omega_{1,2} - \omega_{1,2}/\omega_0)^2 Q^2 = 2$, with the solutions $\omega_1 = \omega_0\left(\sqrt{1+4Q^2} - 1\right)/2Q$ and $\omega_1 = \omega_0\left(\sqrt{1+4Q^2} + 1\right)/2Q$.

The difference $\omega_2 - \omega_1 = \Delta\omega$ defines the *band-width*, i.e. the frequency interval where the amplitude is within -3 dB with respect to the peak value $A(\omega_0)$: $20 \log_{10}(1/\sqrt{2}) \approx -3$.

Comparing filters of *first* and *second order* we see that (in the region $\omega \gg \omega_0$ for the low-pass and in the region $\omega \ll \omega_0$ for the high-pass) $A_{1lp}(\omega) \approx G(\omega_0/\omega)$, $A_{2lp}(\omega) \approx G(\omega_0^2/\omega^2)$, and $A_{1hp}(\omega) \approx G(\omega/\omega_0)$, $A_{2hp}(\omega) \approx G(\omega^2/\omega_0^2)$, respectively. The *order number* measures the steepness of the slope of $A(\omega)$.

In the Bode plot (figure 8.9), where the amplitude is given in dB ($20 \log_{10} A$), in a log-log plot, the slope of a *first-order* low-pass filter (for $\omega \gg \omega_0$) is -20 dB/decade while the slope of a *second-order* low-pass filter is -40 dB/decade. For high-pass filters the slope is (for $\omega \ll \omega_0$) $+20$ dB/decade and $+40$ dB/decade, respectively.

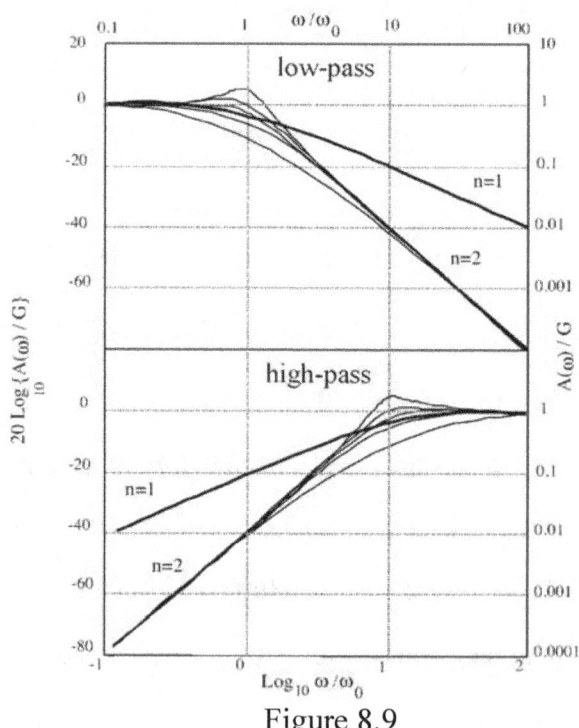

Figure 8.9

8.5. Filters VCVS

A second important group of active filters of second order is the VCVS (*Voltage Controlled Voltage Source*)[35].

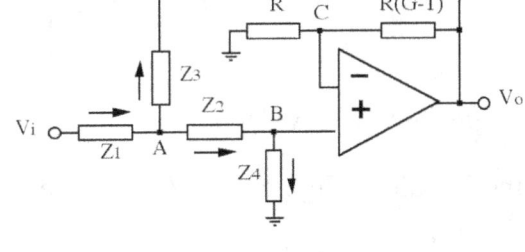

Figure 8.10

These filters are made by one OA ad a passive network with impedances Z_i (R and C) in the general layout of Figure 8.10.

The transfer function may be easily obtained by imposing current conservation at the nodes A ($I_1 = I_2 + I_3$) and B ($I_2 = I_4$), and noting that the voltage is the same at node B and C.

Moreover, the basic non-inverting amplifier gives $V_o = GV_B$.

Equation $I_1 = I_2 + I_3$ at node A gives:

$$\frac{V_i - V_A}{Z_1} = \frac{V_A - V_o}{Z_3} + \frac{V_A - V_B}{Z_2},$$

and relation $I_2 = I_4$ at node B gives:

$$\frac{V_A - V_B}{Z_2} = -\frac{V_B}{Z_4},$$

Letting $V_B = V_o/G$, and solving for V_A yields $V_A = (V_o/G)(1+Z_2/Z_4)$, that, inserted into the first equation gives:

$$\frac{V_i}{Z_i} = -\frac{V_o}{G}\left[\left(1+\frac{Z_2}{Z_4}\right)\left(\frac{1}{Z_1}+\frac{1}{Z_2}+\frac{1}{Z_3}\right)-\frac{G}{Z_3}-\frac{1}{Z_2}\right].$$

The transfer function $T(s) = V_o/V_i$ therefore is:

$$T(s) = -\frac{G}{1+(Z_1/Z_2)/(Z_3/Z_4)+(Z_1+Z_2)/Z_4+(1-G)(Z_1/Z_3)}. \qquad [8.7]$$

The particular choice $G = 1$, transforms the non-inverting amplifier into a buffer and the general layout becomes the simpler one, also named *Sallen-Key filter*[36], shown in Figure 8.11:

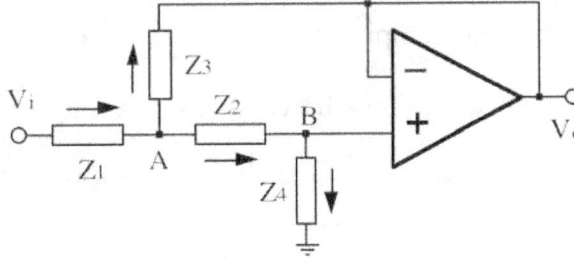

Figure 8.11

---

[35] The name VCVS has historical reason, is is normally used to distinguish this layout, which also has *multiple feedback*, from the previous one: the difference is that now feedback is both positive and negative.

[36] Sallen, R. P.; E. L. Key (1955-03). "A Practical Method of Designing RC Active Filters". *IRE Transactions on Circuit Theory* **2** (1): 74–85.

Which involves only 4 impedances, and the general transfer function becomes:

$$T(s) = \frac{-1}{1+(Z_1/Z_2)/(Z_3/Z_4)+(Z_1+Z_2)/Z_4} \qquad [8.8]$$

8.5.1. The low-pass VCVS

If in the circuit of Figure 8.11 Z_1, Z_2 are resistors and Z_3, Z_4 are capacitors we get a low-pass filter (figure 8.12), with :

$$T(s) = \frac{1}{1+s(R_1+R_2)C_4+s^2(R_1R_2C_3C_4)} = \frac{\omega_0^2}{s^2+s2\zeta\omega_0+\omega_0^2}. \qquad [8.9]$$

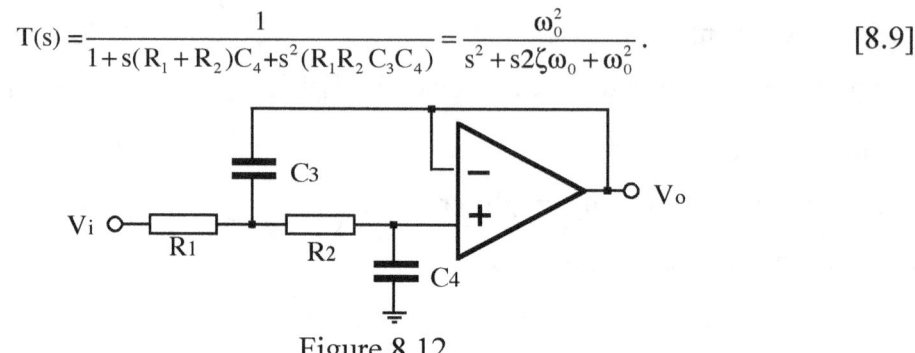

Figure 8.12

The transfer function is similar to that of the multiple feedback low-pass filter of [8.4], but with cut frequency $\omega_0 = 1/\sqrt{R_1R_2C_3C_4}$, and damping factor $\zeta = \tfrac{1}{2}\sqrt{C_4/C_3}\left(\sqrt{R_1/R_2}+\sqrt{R_2/R_1}\right)$.

We may change ω_0, at constant ζ, scaling the resistors by the same factor, or change ζ, at constant ω_0, changing the capacitors while keeping constant their product.

8.5.2. High-pass VCVS

If in the circuit of Figure 8.11 Z_1, Z_2 are capacitors and Z_3, Z_4 are resistors we get a high-pass filter (figure 8.13), with :

$$T(s) = \frac{-s^2(R_3R_4C_1C_2)}{1+sR_3(C_1+C_2)+s^2(R_3R_4C_1C_2)} = \frac{-s^2}{s^2+s2\zeta\omega_0+\omega_0^2}, \qquad [8.10]$$

The transfer function is similar to that of the multiple feedback low-pass filter of [8.5], but with cut frequency $\omega_0 = 1/\sqrt{C_1C_2R_3R_4}$, and damping factor $\zeta = \tfrac{1}{2}\sqrt{R_3/R_4}\left(\sqrt{C_1/C_2}+\sqrt{C_2/C_1}\right)$.

We may change ω_0, at constant ζ, scaling the capacitors by the same factor, or change ζ, at constant ω_0, changing the resistors while keeping constant their product.

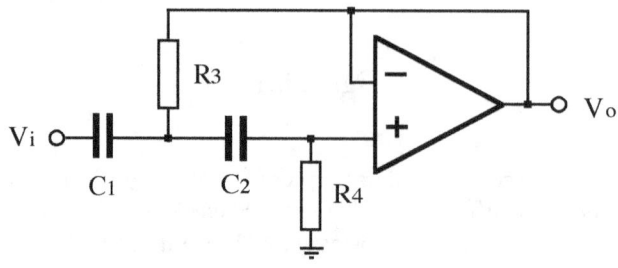

Figure 8.13

If we let G> in the circuit of figure 8.10 the transfer function is simply multiplied by G, and the damping factor ζ becomes respectively:

$$\zeta = \tfrac{1}{2}\sqrt{C_4/C_3}\left(\sqrt{R_1/R_2} + \sqrt{R_2/R_1} + (1-G)\sqrt{(R_1C_3)/(R_2C_4)}\right)$$

for the low-pass,

$$\zeta = \tfrac{1}{2}\sqrt{R_3/R_4}\left(\sqrt{C_1/C_2} + \sqrt{C_2/C_1} + (1-G)\sqrt{(R_3C_2)/(R_4C_1)}\right)$$

for the high-pass.

8.6. The state-variable filters

The state-variable active filters are made of two cascaded inverting integrators plus a summer that adds the outputs of the two integrators (figure 8.14).

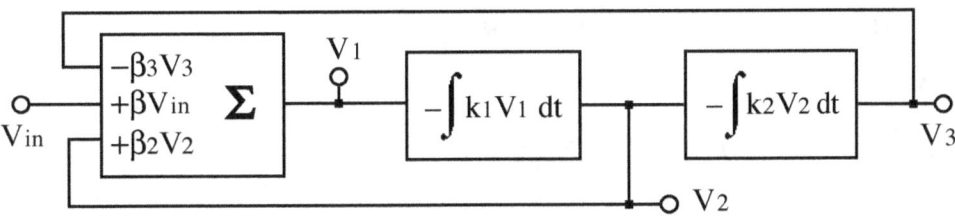

Figure 8.14

To explain the working principle of this kind of filters we start with an example.
In Figure 8.15 we first neglect OA4, which does not affect the behavior of the circuit..
We calculate the voltage V_1 considering that OA1 acts as inverting amplifier for the source V_3, and as non-inverting amplifier with gain 2 for sources V_i and V_2; using the superposition principle we get :

$$V_1 = 2\,[V_iR_2/(R_1+R_2) + V_2R_1/(R_1+R_2)] - V_3.$$

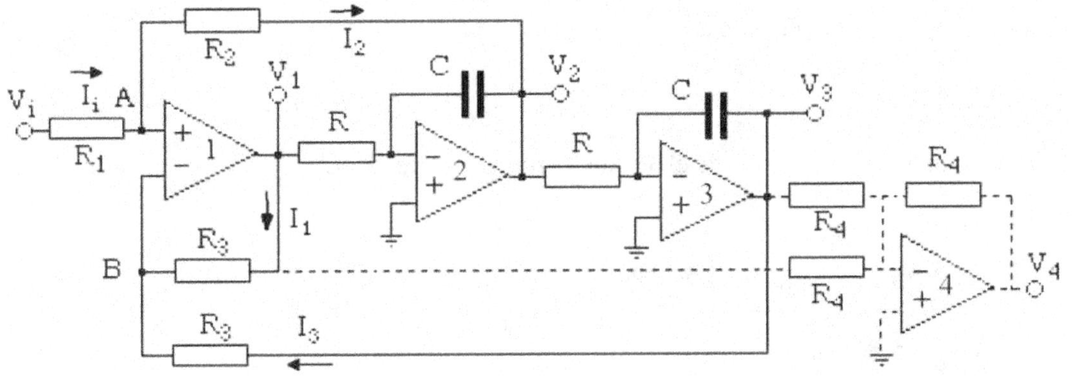

Figure 8.15

The same result may be obtained by applying the conservation of current at nodes A ($I_i = I_2$) and B ($I_1 + I_3 = 0$), and noting that $V_A = V_B$.

Because Oa2 and OA3 are inverting integrators we get $V_2 = -V_1/sRC$ and $V_3 = -V_2/sRC$; by inserting these into the previous equation we obtain:

$$V_1 (1+2R_1/sRC(R_1+R_2)+1/(sRC)^2) = V_i 2R_2/(R_1+R_2).$$

The transfer function at the first output V_1 is:

$$T_1(s) = \frac{V_1}{V_i} = \frac{-s^2 2R_2/(R_1+R_2)}{s^2 + s2R_1/(R_1+R_2)RC + 1/(RC)^2} = \frac{s^2 G_1}{s^2 + s2\zeta\omega_0 + \omega_0^2}, \qquad [8.11]$$

where $\omega_0 = 1/RC$, $\zeta = R_1/(R_1+R_2)$ and $G_1 = 2R_2/(R_1+R_2)$

The transfer functions for the other two outputs V_2 and V_3 are therefore:

$$T_2(s) = \frac{V_2}{V_i} = -\frac{V_1 \omega_0}{V_i s} = \frac{-sG_1 \omega_0}{s^2 + s2\zeta\omega_0 + \omega_0^2} = \frac{sG_2 \omega_0/Q}{s^2 + s\omega_0/Q + \omega_0^2} \qquad [8.12]$$

where $Q = 1/2\zeta = (R_1+R_2)/2R_1$ and $G_2 = QG_1 = R_2/R_1$, and

$$T_3(s) = \frac{V_3}{V_i} = \frac{V_1 \omega_0^2}{V_i s^2} = \frac{G_1 \omega_0^2}{s^2 + s2\zeta\omega_0 + \omega_0^2}. \qquad [8.13]$$

Comparing [8.11], [8.12], [8.12], with [8.4], [8.5], [8.6], we see immediately that at V_1, V_2, V_3 we have a high-pass, a band-pass and a low-pass.

Considering now also OA4 (an inverting summer for V_1 and V_3) we obtain at the fourth output:
$V_4 = -(V_1 + V_3)$, with the transfer function:

$$T_4(s) = \frac{V_4}{V_i} = \frac{-G_1(s^2 + \omega_0^2)}{s^2 + s\omega_0/Q + \omega_0^2}. \qquad [8.14]$$

Relation [8.14] describe the behavior of a *band-reject* (or notch) filter: for $s^2 >> \omega_0^2$ or $s^2 << \omega_0^2$ the amplitude $A(s) \rightarrow G$, while for $\omega = \omega_0$, $A(s) = 0$.

The band-width is $\Delta\omega = \omega_0/Q = 2R_1/RC(R_1+R_2)$, the same as that of the band-pass filter.

Note that the state-variable filters are devices that may be used as analogic computers to solve differential equations.

For example in Figure 8.14, because $V_3 = -k_2 \int V_2 \, dt$, we have $V_2 = -(1/k_2) \partial V_3/\partial t$, and also $V_1 = -(1/k_1)\partial V_2/\partial t = (1/k_1 k_2)\partial^2 V_3/\partial t$.

Letting $V_3 = y(t)$, $1/k_1 k_2 = a$, $-\beta_2/k_2 = b$, $\beta_3 = c$ and $-\beta V_i = d$, the function $y(t)$ satisfies the differential equation $a\partial^2 y/\partial t + b\partial y/\partial t + cy + d = 0$.

This result is general: for any linear differential equation we may find a circuit, made of integrators and summers, which gives the solving function..

8.7. A simple notch filter

A *notch filter* may also be made of a single AO, as shown in Figure 8.16. This circuit may be seen as a modification of that shown in figure 8.3, by letting $Z_2=\infty$ and by feeding a fraction G of the input signal to the non-inverting input.

Note that in this circuit the values of capacitors and resistors are not arbitrary! In fact we must set $Z_4=Z_3=1/sC$ and $G=R_2/(R_2+2R_1)$.

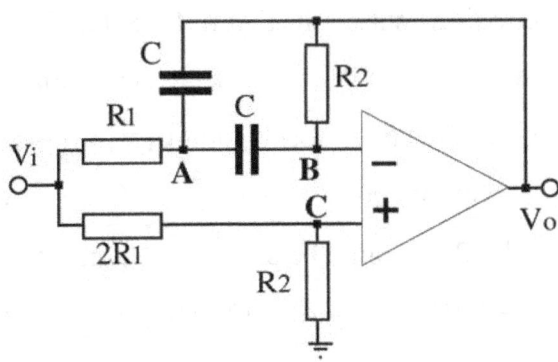

Figure 8.16

The transfer function may be calculated by imposing the current conservation at nodes A: $(V_i-V_A)/R_1=(V_A-V_o)sC+(V_A-V_B)sC$, and at node B: $(V_A-V_B)sC=(V_B-V_o)/R_2$, and by noting that $V_B=V_C=V_iR_2/(R_2+2R_1)$.

$$T(s) = \frac{G(s^2+1/R_1R_2C^2)}{s^2+2s/R_2C+1/R_1R_2C^2} = \frac{G(s^2+\omega_0^2)}{s^2+s\omega_0/Q+\omega_0^2} \quad [8.15]$$

Here $\omega_0=1/\sqrt{R_1R_2}C$, $Q=\tfrac{1}{2}\sqrt{R_2/R_1}$, $G=R_2/(R_2+2R_1)<1$, and the band-width is $\Delta\omega=\omega_0/Q=2/R_2C$.

8.8. The impedance converter (NIC)

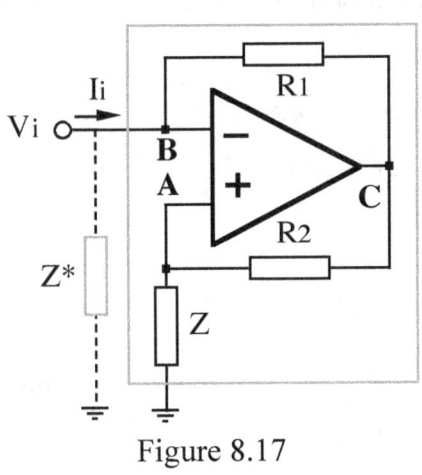

Figure 8.17

The circuit of Figure 8.17, converts the impedance Z into the impedance $Z^*=-(R_1/R_2)Z$ (a *negative impedance*).

By definition, the input impedance is $Z^*=V_i/I_i$, and the input current may be written $I_i=(V_i-V_C)/R_1$. On the other hand the output voltage V_C may be calculated as:

$$V_C = -AV_B + AV_A = -AV_i + AV_C[Z/(R_2+Z)]$$

where A is the open-loop gain, which gives:

$$V_C = -V_i / \left(\frac{1}{A} - \frac{Z}{R_2+Z}\right) \rightarrow V_i\left(1+\frac{R_2}{Z}\right)$$

Putting all together we get $Z^*=V_i/I_i=V_iR_1/(V_i-V_C)=-ZR_1/R_2$.

8.8.1 A band-pass NIC filter

We may obtain a band-pass filter using a NIC circuit as shown in Figure 8.18 The ideal OA model gives $V_C=V_A$, and the Ohm's Law gives $I_A = (V_A-V_B)/R_1$ and $I_B = (V_B-V_C)/R_2$. Therefore $I_B = -I_A R_1/R_2 = -G^* I_B$, where $G^* = R_1/R_2$ may be seen as the current gain of the NIC of Figure 8.17. Moreover $V_o = Z_b I_B$ (with $Z_b = R_b \| 1/sC_b$) and $V_i - V_A = Z_a I_A$ (with $Z_a = R_a + 1/sC_a$) and $V_o = V_C$.

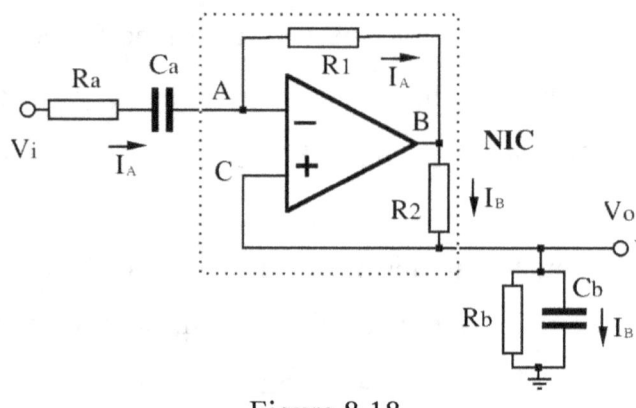

Figure 8.18

By solving the system of all the above equations we obtain the transfer function V_o/V_i:

$$T(s) = \frac{-sG^*/(R_a C_b)}{s^2 + s(1/R_a C_a + 1/R_b C_b - G^*/R_a C_b) + 1/R_a R_b C_a C_b} = \frac{-sG\omega_0/Q}{s^2 + s\omega_0/Q + \omega_0^2} \qquad [8.16]$$

By comparing [8.16] with [8.6], we see that this is a band-pass filter, with central frequency $\omega_0 = 1/\sqrt{R_a R_b C_a C_b}$, gain $G = G^*/(C_a/C_b + R_a/R_b - G^*)$.

The Q-factor $Q = 1/\left(\sqrt{R_b C_b/R_a C_a} + \sqrt{R_a C_a/R_b C_b} - G^*\sqrt{R_b C_a/R_a C_b}\right)$ may be adjusted by simply changing the ratio G^* of the resistors R_1, R_2. The G^* value, however is not arbitrary: we must avoid excessive G values. E.g. for $R_a = R_b = R$ and $C_a = C_b = C$, we get $G = G^*/(2-G^*)$ and $Q = 1/(2-G^*)$. Both $Q \to \infty$ and $G \to \infty$ for $G^* \to 2$, so that the filter stop working for G^* values too close to 2, because the OA saturates.

An equivalent method to derive T(s), is to use the result obtained in § 8.8.
Then $T(s) = V_o/V_i = Z^*/(Z_i + Z^*) = G^* Z_b/(Z_a + G^* Z_b)$, that gives again [8.16].

8.9. Gyrator

The gyrator [37] is a circuit that converts an impedance into its reciprocal, scaled by a factor K: $Z^* = K/Z$. If Z is a capacitor ($Z = 1/sC$), the effective impedance seen from the gyrator input is $Z^* = sKC$, equivalent to the inductance $L^* = KC$. An example is the circuit of Figure 8.19.

By definition the input impedance is $Z_i = V_i/I_i$. The negative feedback gives : $V_A = V_B = V_i$ and the Ohm's Law gives: $I_i = (V_i - V_2)/R_1$, so we only need to calculate V_2.

---

[37] The name explains that it *rotates* the vector associated to the complex impedance (changing a capacitor into an inductance the Z-vector rotates by 180°).

At the node A : $(V_1-V_A)/R_3 = (V_A-V_2)/R_2$, that solved for V_2 gives:

$V_2 = (1+R_2/R_3)V_i - R_2/R_3 V_1$.

At the node B : $V_B = V_1 R_5/(R_5+Z_4)$, that solved for V_1 gives $V_1 = V_i(1+Z_4/R_5)$.

Therefore $Z_i = R_1 R_3 R_5/(R_2 Z_4) = sR_1 R_3 R_5 C/R_2 = L^*$, where $L^* = R_1 R_3 R_5 C/R_2$ (the effective inductance) is the capacitance multiplied by $R_1 R_3 R_5/R_2$.

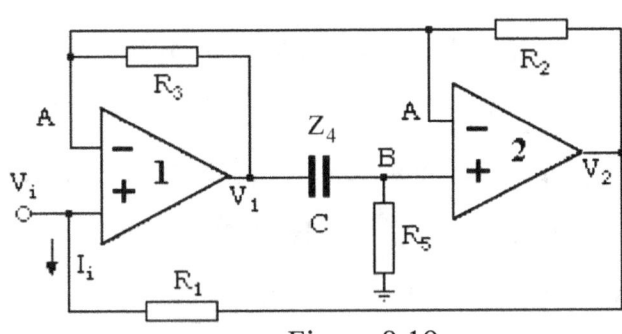

Figure 8.19

This circuit is equivalent to an inductance whose value may be made quite large, useful for obtaining low-pass LC filters with very low cut frequency[38]. E.g. with 1 kΩ resistors, we get $L^*/C = 1$ henry/µF.

There are also commercial IC (integrated circuits), like the National AF120, that make easy setting-up the gyrator (Figure 8.20).

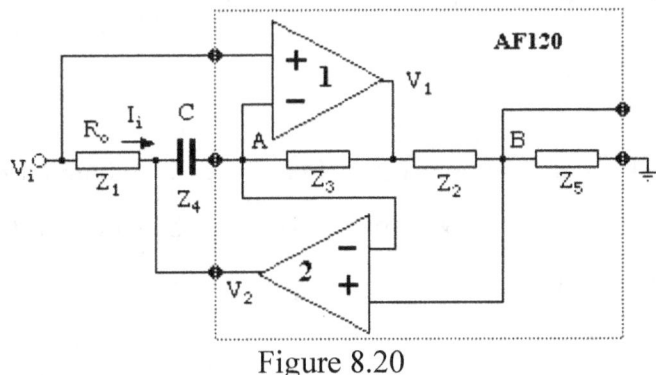

Figure 8.20

Circuits 8.19 and 8.20 differ only for the position of C, which in 8.20 is exchanged with R_2. The previous analysis gives $Z_i = Z_1 Z_3 Z_5/(Z_2 Z_4)$, with $Z_2 = 1/sC$. Integrated in AF120 there are $Z_3 = Z_2 = Z_5 = R = 7.5$ kΩ, so that letting $Z_1 = R_o$ we obtain $Z_i = sRR_o C = sL^*$, with $L^*/R_o C = 7.5$ (henry/ms).

8.10. Capacitance multiplier

The circuit shown in Figure 8.21 behaves as a capacitance multiplier.

To calculate the input impedance $Z_i = V_i/I_i$ we must evaluate the input current $I_i = (V_i-V_2)/Z_C$. Because OA1 is a follower we have $V_1 = V_i$, and because OA2 is an inverting amplifier with gain $G = -R_o/R_i$, we get $V_2 = GV_1$.

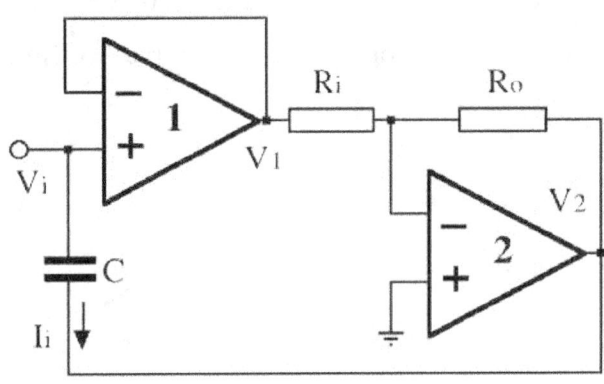

Figure 8.21

---

[38] See the examples of LC filters in Appendix B.

As a conclusion: $I_i = V_i(1+R_o/R_i)sC$, and $Z_i = 1/sC^*$, where the effective capacitance is $C^* = C(1+R_o/R_i)$. This circuit may be used, with a capacitor R in series to the input, as an RC* low-pass filter, with the output taken at the node between R and C*.

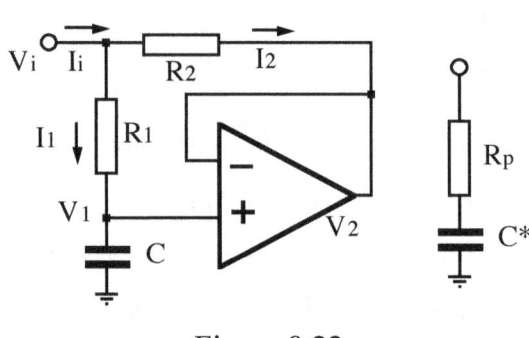

Figure 8.22

Another capacity multiplier is shown in Figure 8.22, where the OA is used as follower, so that $V_1 = V_2$.

The voltage V_1 is calculated from the divider $(R_1, Z_C) : V_1 = V_i Z_C/(R_1+Z_C) = V_i/(1+sR_1C)$.

The input impedance $Z_i = V_i/I_i$ is calculated considering that $I_i = I_1 + I_2$, that gives $I_i = (V_i-V_2)/R_1 + (V_i-V_2)/R_2 = (V_i - V_2)/R_p$, where $R_p = R_1 \| R_2$. Eliminating V_2 we obtain $Z_i = R_p(1+1/sR_1C) = R_p + 1/sC^*$, where C* is the effective capacity $C^* = C(1+R_1/R_2)$.

8.11. IC active filters

The state-variable filters may be easily obtained using commercially available as IC. A typical example is the National AF100, (or the similar Intersil FLTU2), whose internal structure is shown in Figure 8.23.

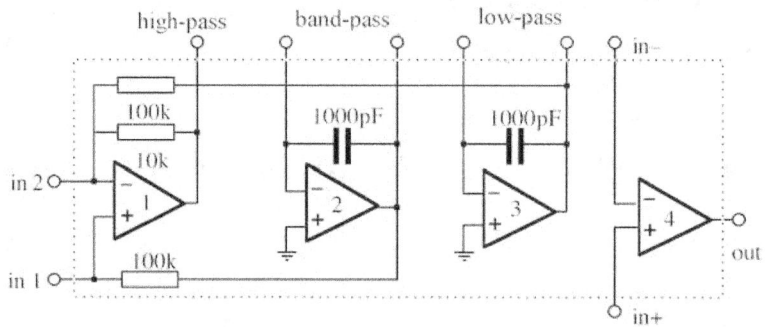

Figure 8.23

A possible configuration of AF100 is shown in Figure 8.24, where, ignoring the fourth OA of figure 8.23, we obtain the same circuit of figure 8.15.

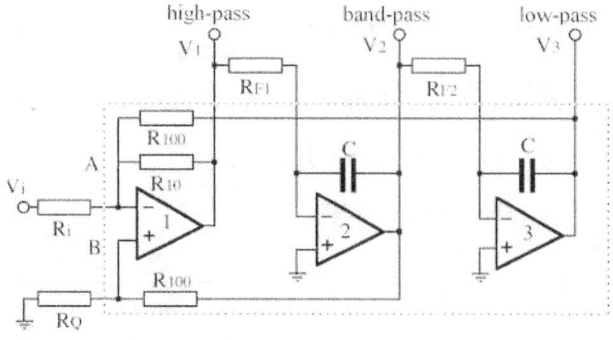

Figure 8.24

The current conservation at node A gives: $(V_A-V_i)/R_i=(V_3-V_A)/R_{100}+(V_1-V_A)/R_{10}$,

and at node B: $(V_2-V_A)/R_{100}=V_A/R_Q$.

From integrator OA1 we get: $V_2=-V_1/sR_{F1}C=-\omega_1 V_1/s$ and from integrator OA2: $V_3=V_2/sR_{F2}C=\omega_1\omega_2 V_1/s^2$, where $\omega_1=1/R_{F1}C$ and $\omega_2=1/R_{F2}C$. Therefore:

$$V_A = -(R_Q\|R_{100})\omega_1 V_1/s.$$

Eliminating V_a and V_3 in the first equation we get the transfer function for V_1 (high-pass):

$$T_1(s) = \frac{-s^2 R_{10}/R_i}{s^2 + s\omega_1 R_{10}(R_Q\|R_{100})/[R_{100}(R_i\|R_{10}\|R_{100})] + \omega_1\omega_2 R_{10}/R_{100}}.$$

The cut frequency (with C=1nF, R_{10}=10 kΩ, R_{100}=100 kΩ) is:

$$\omega_0 = \sqrt{\omega_1\omega_2/10} = 10^9\sqrt{1/(10\,R_{F1}R_{F2})},$$

e.g. for R_{F1}=10 kΩ, R_{F2}=1 kΩ, ω_0=100kHz; the damping factor ζ is:

$$\zeta = \frac{1}{2}\sqrt{\frac{10 R_{F2}}{R_{F1}}}\,\frac{1.1+10^4/R_i}{1+10^5/R_Q}$$

e.g. for R_i=10 kΩ, R_Q=1kΩ, $\zeta \approx 0.5$ %; and the gain is $G_1=R_{10}/R_i$.

In the band-pass we get $G_2 = Q(R_{10}/R_i)\sqrt{10 R_{F1}/R_{F2}}$, with a quality factor $Q=1/(2\zeta)$, in the low-pass the gain is $G_3=R_{100}/R_i$.

Note that gain and quality factors may be varied at constant ω_0 by properly adjusting R_Q and R_i.

With AF100 we need only four resistors to get a triple filter. With three more resistors (and using the fourth AO of AF100) we may build the notch filter of figure 8.15, as shown in figure 8.25.

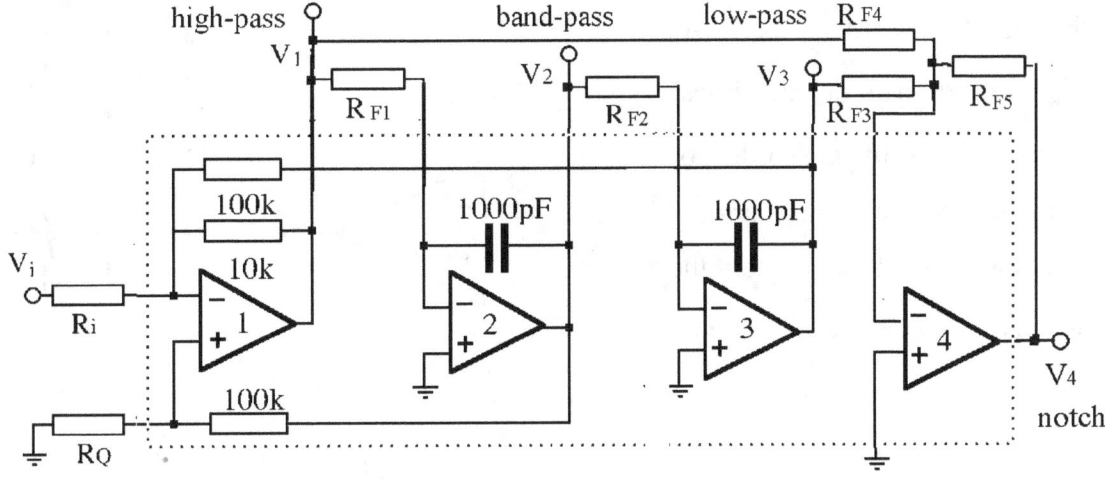

Figure 8.25

9. Switching circuits

When the OA has no negative feedback, or it has a large positive feedback, a small noise voltage at the input (e.g. offset, line pick-up, switch-on transients...) brings the output to saturation. The OA works out of the linear region and its response to input voltage may take only two values: V_{cc}^+ or V_{cc}^-.[39]

This behavior allows us to use the OA as a switching circuit, i.e. as a *comparator*. Not all the commercial OA may be used for this purpose: many models suffer of *latch-up*, i.e. they get blocked with output saturated, and to unlatch them we must switch-off the power supply.

Therefore, when designing a switching circuit we must select special OA with *rail-to-rail* output, that do not suffer latch-up, named *Schmitt triggers* or *Comparators*. Some comparator are available with *open-collector* [40], a configuration that allows to select for saturation voltage (V_o) values different from the power supply voltages.

9.1. Comparator

Let us first analyze an OA without negative feedback. We immediately see that it works as a threshold detector. In fact if we fix one of the inputs at a reference voltage V_R, the output switches between $\pm V_{cc}$ as soon as the voltage applied to the other input crosses the threshold voltage V_R. For example, let $V_R > 0$, the output voltage V_o, as a function of the input voltage V_i, is shown in Figure 9.1. Within the small range $\Delta V = 2V_{cc}/A_o$ around V_R the comparator has linear response, but ΔV is of the order of millivolt, so that a small noise around V_R makes the output unstable: the comparator oscillates between $+V_{cc}$ and $-V_{cc}$.

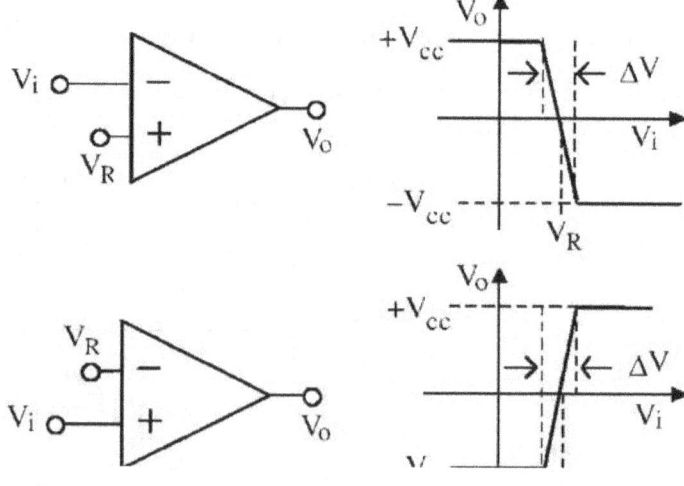

---

[39] Here we assume for simplicity $V_{cc}^+ = -V_{cc}^-$, and $|V_{oMax}| \approx V_{cc}$.
[40] See Figure 12.8 of chapt. 12 for open-collector layout.

9.2. Comparator with hysteresis

The comparator instability around V_R may be avoided, by introducing an hysteresis through a positive feedback. In this case the response, within a small range around V_R, will depend on the values previously assumed by the input V_i. The single threshold value will be replaced by two threshold values: a lower one, that will switch the output for increasing input voltages, and a higher one, that will switch the output for decreasing input voltages.

Therefore small oscillations of the input voltage V_i nearby each threshold value will not toggle the output more than once. The larger is ΔV, named hysteresis width, the smaller is the comparator sensitivity.

Let us analyze the inverting comparator with $V_R > 0$:

The non-inverting input voltage is set by the superposition of two sources: the output voltage V_o and the reference voltage V_R, as well as by the divider (R_1, R_2), i.e. by the *feedback fraction*. $\beta = R_1/(R_1+R_2)$.

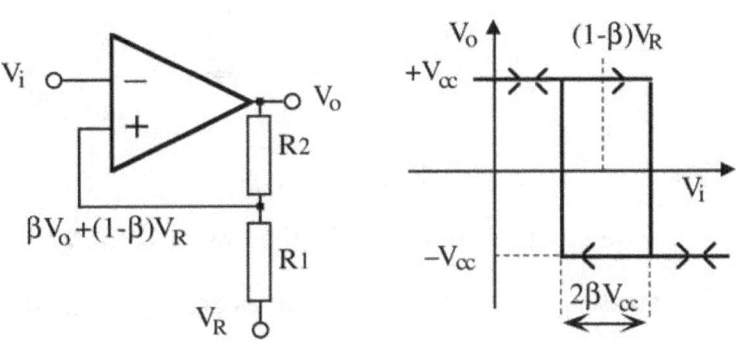

Figure 9.2

The threshold voltages are $\pm\beta V_{cc} + (1-\beta)V_R$. The hysteresis width $2\beta V_{cc}$ replaces the linear region. The mean value of threshold voltages $(1-\beta)V_R$ well approximates V_R for $\beta \ll 1$.

The non-inverting comparator with hysteresis (Figure 9.3) is similar with the difference that the input impedance, $Z_{in} = R_1 \| R_2$ is here lower than Z_{in2}.

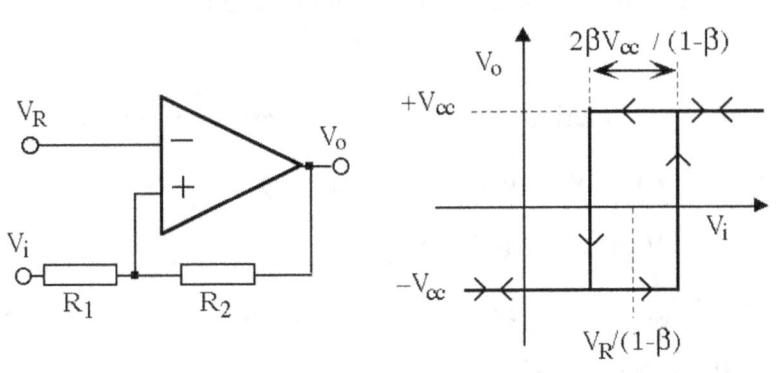

Figure 9.3

This comparator toggles when $\beta V_o + (1-\beta)V_{in} = V_R$ i.e. for $V_{in} = (V_R \pm \beta V_{cc})/(1-\beta) = V_R \pm R_1/R_2 V_{cc}$.

For $\beta \ll 1$ the threshold becomes $V_{in} \approx V_R \pm \beta V_{cc}$ i.e. the hysteresis width is $2\beta V_{cc}/(1-\beta) \approx 2\beta V_{cc}$.

In the particular case $V_R = 0$, the toggling condition is $V_{in} = \pm R_1/R_2 V_{cc}$.

9.3. Bipolar astable multivibrator

If we replace the input signal of an inverting comparator by a complex (RC) negative feedback, we obtain an astable monovibrator, a type of *relaxation oscillator* [41].

We first consider the case of bipolar power supply ($-V_{cc} < V_o < +V_{cc}$), and we let $V_R = 0$, shown in the circuit of Figure 9.4.

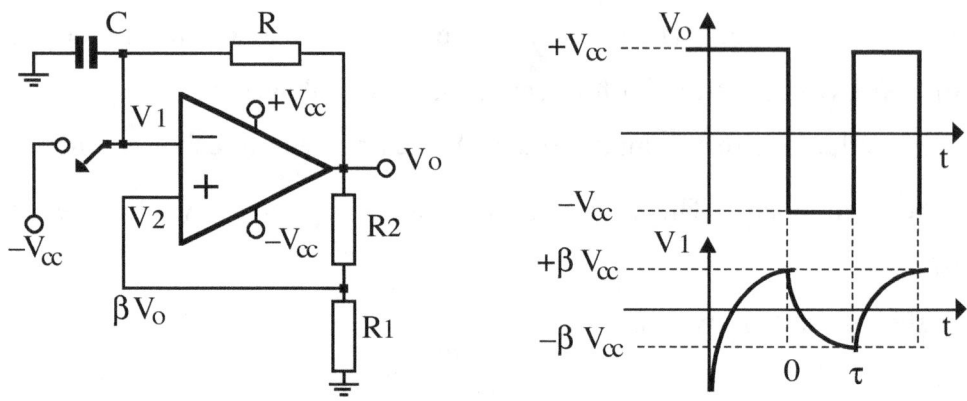

Figure 9.4

The negative feedback forces the voltage V_1 at the inverting input to follow the output voltage V_o, with the delay produced by the low-pass filter RC.

Because $V_R = 0$, the threshold values are $\pm \beta V_o$, and the time evolution of V_1 and V_o are shown in the figure 9.4. Let the switch be initially closed, forcing $V_1 = -V_{cc}$, $V_o = +V_{cc}$ and $V_2 = +\beta V_{cc}$. The capacitor is initially charged, and a current $i = 2V_{cc}/R$ starts discharging it through the resistor R when the switch is opened. When the voltage V_1 reaches $V_2 = +\beta V_{cc}$ the comparator output switches to $V_o = -V_{cc}$ changing the threshold voltage into the new value $V_2 = -\beta V_{cc}$: we assume this instant as $t = 0$.

At this time $V_1(t)$ becomes the exponential function decaying with time constant RC, and boundary conditions : $V_1(0) = +\beta V_{cc}$ and $V_1(\infty) = -V_{cc}$; therefore we may write:

$$V_1(t) = (\beta V_{cc} + V_{cc}) \exp\{-t/RC\} - V_{cc}.$$

When $V_1(t)$ reaches the threshold $V_2 = -\beta V_{cc}$, again the comparator switches to $V_o = +V_{cc}$: we name this time $t = \tau$. The new exponential law becomes:

$$V_1(\tau) = -\beta V_{cc} = (\beta V_{cc} + V_{cc}) \exp\{-\tau/RC\} - V_{cc},$$

from which we get $\tau = RC \ln\{(1+\beta)/(1-\beta)\} = RC \ln\{1+2R_1/R_2\}$.

For $\beta \ll 1$, (i.e. $R_1 \ll R_2$) we have $\tau \approx 2RCR_1/R_2$.

---

[41] See for example http://en.wikipedia.org/wiki/Relaxation_oscillator

At the time t=τ the comparator switches again, the voltage $V_1(t)$ increases again towards $+V_{cc}$ and the threshold is again $V_2=+\beta V_{cc}$. It is easy to see that the next switch of the output voltage occurs after the time τ'=τ: the output signal $V_o(t)$ is therefore a *square-wave* with period T=2τ, or frequency f=1/T=1/2τ. For $\beta \ll 1$ we get $f \approx R_2/4RCR_1$, i.e. a frequency linearly increasing with R_2, or a period linearly increasing with R, C, R_1.

The square-wave symmetry (i.e. τ' = τ) is due to the power supply symmetry ($V^+_{cc} = V^-_{cc}$) and to the choice $V_R=0$. In case of non-symmetrical power supply we may add a double zener in parallel to the output load and a resistor R_o, as shown in Figure 9.5.

When the OA output V'_o reaches V^+_{cc} or V^-_{cc}, the oscillator output V_o is forced to $\pm V_z$.

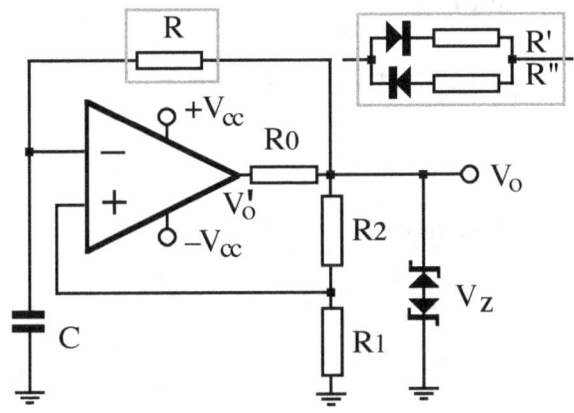

The voltage drop $V^+_{cc} - V_z$ or $V_z - V^-_{cc}$ across the resistor R_o removes the effect of non-symmetric power supply.

If we need a pulser with τ' ≠ τ we may replace the negative feedback resistor R by a parallel of two resistors in series with opposite diodes, (see insert in Figure 9.5).

Figure 9.5

9.4. Unipolar astable multivibrator

If only unipolar power supply is available (V^+_{cc}, 0), or if we need positive output pulses we may use the circuit of Figure 9.6.

In the general case ($R_1 \neq R_2 \neq R_3$) the voltage V_2 takes the threshold voltages V_2^+ e V_2^- for output $V_o = V_{cc}$ or $V_o = 0$:

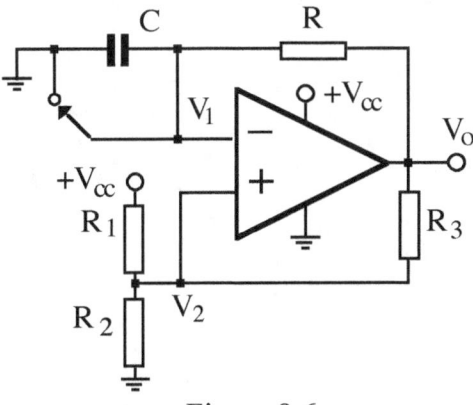

Figure 9.6

$$V_2^+ = V_{cc} \frac{R_2}{R_2 + R_1 \| R_3} = V_{cc} \frac{R_2(R_1 + R_3)}{R_1 R_2 + R_2 R_3 + R_1 R_3}$$

$$V_2^- = V_{cc} \frac{R_2 \| R_3}{R_1 + R_2 \| R_3} = V_{cc} \frac{R_2 R_3}{R_1 R_2 + R_2 R_3 + R_1 R_3}$$

In the simpler case $R_1=R_2=R_3=R$ we get $V_2^+ = \tfrac{2}{3} V_{cc}$ e $V_2^- = \tfrac{1}{3} V_{cc}$.

With close switched: $V_1 = 0$ and $V_o = V_{cc}$. When the switch is opened, the capacitor starts to be charged through R and the voltage V_1 grows, crossing the threshold V_2^+ at the time that we assume to be t = 0. The comparator toggles forcing $V_0 = 0$ and $V_2 = V_2^-$.

The time evolution for V_1 becomes $V_1(t) = V_2^+ \exp(-t/RC)$. The next comparator toggle occurs at $t = \tau_1$, when V_1 reaches the lower threshold V_2^-: $\exp(-\tau_1/RC) = V_2^-/V_2^+ = R_3/(R_1+R_3)$, that gives:

$$\tau_1 = RC \ln(1 + R_1/R_3).$$

At this time (we again set as t=0) $V_o = V_{cc}$, and the voltage V_1, starting from V_2^-, grows toward V_{cc} and the next comparator toggling occurs at $t = \tau_2$. The time evolution is now $V_1(t) = (V_2^- - V_{cc})\exp(-t/RC) + V_{cc}$, that gives for the positive pulse width τ_2:

$$\tau_2 = RC \ln[(V_2^- - V_{cc})/(V_2^+ - V_{cc})] = RC \ln(1 + R_2/R_3).$$

For $R_1 = R_2$ we get a square-wave ($\tau_2 = \tau_1$).

The same circuit, with power supply $(0, -V_{cc})$ gives negative pulses.

10. Self-oscillation

Self-oscillation in OA is a spontaneous oscillation of the output voltage in the absence of input signal: it may occur when there is a *positive feedback*.

Positive feedback may be provided by a fraction of the output signal fed to the *non-inverting* OA input, but also a fraction of the output signal fed to the *inverting* OA input, if there is a *phase shift of* π.

Such positive feedback may also be non-intentional: it may be the result of capacitive coupling between output and input or it may be due to a ground-loop[42] in the power supply circuitry; in these cases the oscillation is undesired, not controlled and it produces instability of the signals.

If we properly adjust the positive feedback, however, we may obtain stable and controllable oscillation:

10.1. General remarks

Let us consider a loop made by an amplifier with gain A and a total feedback fraction β.

Suppose we inject a signal into any point of the closed-loop: we'll find that signal amplified of the factor $A\beta$ after one loop-turn. Both the gain and the feedback fraction are generally complex function of frequency: $A\beta = A(s)\beta(s)$, with $s=j\omega$.

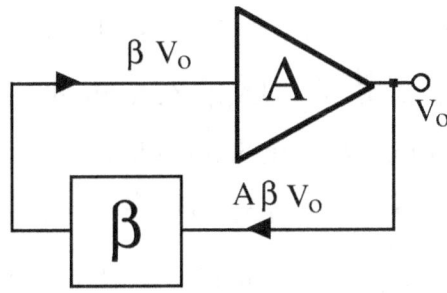

Figure 10.1

Therefore also the *loop-gain* $A\beta$ is a transfer function of the frequency: $A\beta=a(s)+jb(s)$ where a is the *real part* and b is the *imaginary part*; and also we may write $A\beta = a + jb = \sqrt{a^2+b^2}\,e^{j\phi}$ with $\phi = \arctan(b/a)$. So that when the *imaginary part* $b=0$ the *phase shift* ϕ of $A\beta$ is zero and the amplitude gain $|A\beta| = a = $*real part* of $A\beta$.

For $|A(\omega)\beta(\omega)|\geq 1$, and $\phi(\omega)=0$, any noise inside the loop will trigger a signal at frequency ω that will increase with time. If $|A(\omega_0)\beta(\omega_0)|=1$, and $\phi(\omega_0)=0$, the signal at frequency ω_0 stabilizes and this phenomenon is named *self-oscillation*.

Self-oscillation, therefore, is not possible for any frequency: two equations must be satisfied: : $\text{Im}[A(\omega_0)\beta(\omega_0)] = 0$, which means *zero phase shift*, and $\text{Re}[A(\omega_0)\beta(\omega_0)] = 1$, which means that the amplitude loop-gain equals one. In fact for $\text{Re}\{A\beta\}<1$ the oscillation dies-out, and if

---

[42] See for example http://en.wikipedia.org/wiki/Ground_loop_%28electricity%29

Re{$A\beta$}>1 the OA goes to saturation, the signal becomes distorted and the circuit analysis become more complex.

10.2. Wien-bridge sinusoidal oscillator

A simple example of sinusoidal oscillator, named *Wien-bridge oscillator*, is shown in Figure 10.2.

Within the ideal OA model, gain is $A = 1+R_o/R_1$: it is a real number, i.e. it does not depend on ω.

The feedback β fraction, instead, is a complex function of ω (the low-pass filter transfer function).

From relation $\beta = (Z_{C3} \| R_3)/(Z_{C3} \| R_3 + R_2 + Z_{C2})$ we get

$$\beta(j\omega) = \frac{j\omega R_3 C_2}{(j\omega/\omega_0)^2 + j\omega/\omega_0 Q + 1}, \text{ with}$$

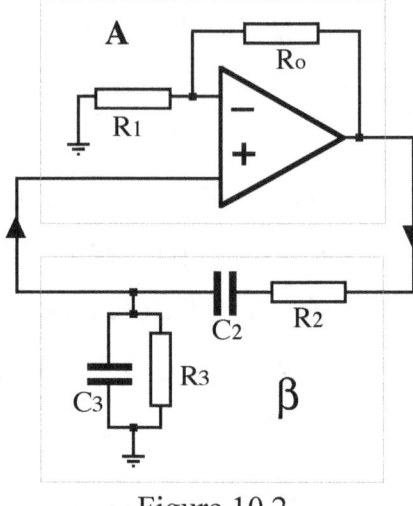

Figure 10.2

$\omega_0 = \sqrt{1/(R_2 R_3 C_2 C_3)}$ and $Q = 1/[\omega_0 (R_3 C_2 + R_2 C_2 + R_3 C_3)]$.

The condition $\text{Im}[A\beta]=0$, with A real, becomes $\text{Im}[\beta]=0$, an equation that is easily solved noting that the solution is obtained by imposing zero real part in the denominator of β (because the numerator is imaginary), and, noting that $(j\omega/\omega_0)^2 = -(\omega/\omega_0)^2$, finally we obtain the solution: $\omega = \omega_0$. For $\omega = \omega_0$ the feedback fraction becomes:

$$\beta(\omega_0) = R_3 C_2 \omega_0 Q = 1/(1 + R_2/R_3 + C_3/C_2) \,. \tag{10.1}$$

The oscillation become stable for $|A\beta|=1$, i.e. when $R_o/R_1 = R_2/R_3 + C_3/C_2$.

The simple case is for $R_2=R_3=R$, $C_3=C_2=C$, corresponding to $\beta=1/3$ and $Q=1/3$, imposes $A=3$, i.e., $R_o=2R_1$ so that the oscillation frequency is $f_o = 1/(2\pi RC)$.

Another simple choice is $R_3=2R_2$ and $C_2=2C_3$ (i.e. $\beta=1/2$, $Q=1/4$) imposes $A=2$, i.e. $R_o=R_1$. We are free in setting the values in the feedback network, provided that we satisfy the conditions [10.1] and $|A\beta|=1$. At high frequencies we must account for the frequency dependence of the open-loop gain $A_{ol}(\omega)$, which decreases with ω.

A similar circuit may be obtained by replacing the capacitors in Figure 10.2 with inductances. We would get $1/\beta = 1+R_2/R_3+L_2/L_3+j(\omega L_2/R_3 - L_2/R_3\omega)$, $(\omega_0)^2 = (R_2 R_3/L_2 L_3)$; for $R=R_2=R_3$, $L=L_2=L_3$ the oscillation frequency would be $f_o = R/2\pi L$.

The closed-loop gain value is critical: it must be *exactly* $A\beta|=1$. Therefore a stable oscillator normally requires an automatic gain stabilization (note that the amplitude of the voltage oscillation does not enter explicitly into the equations we used above).

This may be achieved using *non-linear* passive elements in the feedback network, as in the circuits of Figure 10.3.

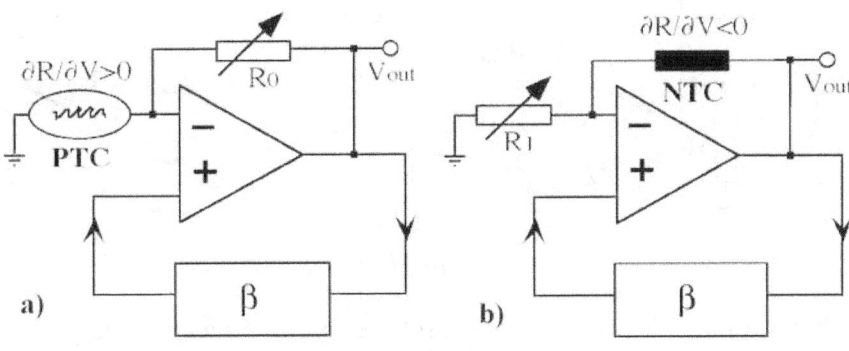

Figure 10.3

In both cases we exploit the temperature dependence of a non-linear resistor R(T): because the power $W=V^2/R$ dissipated on R(T) increases with the voltage amplitude, also the temperature increases. In the case a) of PTC (Positive Temperature Coefficient) thermistor (it might be simply a filament lamp) the result is $\partial R/\partial V>0$, so that the loop-gain $A=1+R_o/R(V)$ decreases.

In the case b) with NTC (Negative Temperature Coefficient) thermistor we have $\partial R/\partial V<0$, and the gain is $A=1+R(V)/R_1$.

Assuming for example $\beta = 1/3$ must be $A = 3$: therefore, at room temperature we should choose for the PTC : $R_o=2R_{PTC}$ and for the NTC : $R_1=R_{NTC}/2$.

Another Wien-bridge oscillator circuit is shown in Figure 10.4. Here the automatic gain control is provided by the non-linear behavior of the diodes placed in parallel to R_f. At higher oscillation voltages the diodes start conducting, thus decreasing the effective feedback resistance, and the closed-loop gain (that is initially set to A=3 by adjusting the potentiometer R).

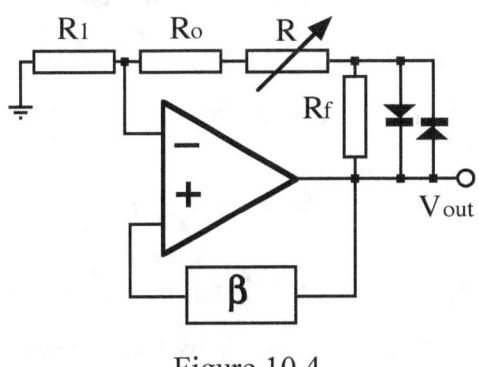

Figure 10.4

10.3. Phase shifter

A phase shifter is an all-pass filter, that does not affects the signal amplitude $|T(j\omega)|=1$, while introducing a phase shift that does depend on frequency.

Two examples are shown in Figure 10.5.

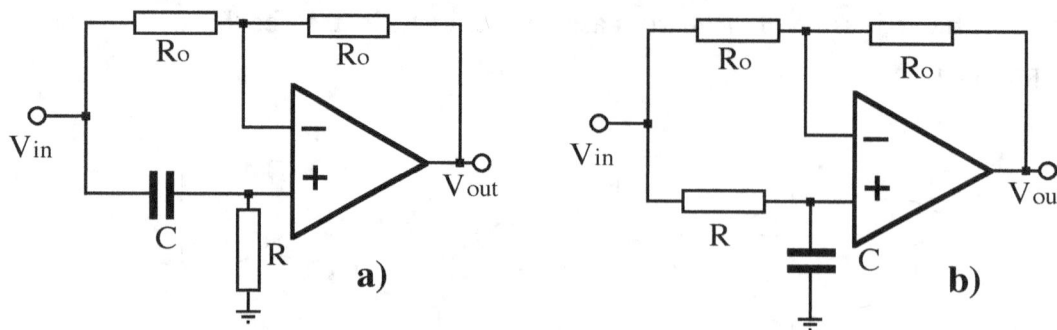

Figure 10.5

The circuit a) produces a *negative* phase offset and the circuit b) a *positive* phase offset: note that the two circuits differ only for the position of R and C. We analyze both, redrawing the circuit in the general layout of Figure 10.6. Voltage V_{out} is the superposition of the source V_{in} amplified -1 (inverting) and of the source $Z_2/(Z_1+Z_2)V_{in}$ amplified $+2$ (non-inverting).

The transfer function is:

$$T(j\omega) = \frac{Z_2 Z_1}{Z_1 + Z_2} - 1 = \pm \frac{1 - j\omega RC}{1 + j\omega RC} = \pm e^{-j2\phi},$$

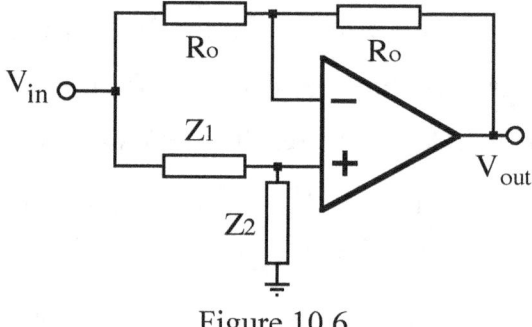

Figure 10.6

with the sign $-$ in case a) and sign $+$ in case b). The phase shift is $2\phi = 2\arctan(\omega RC)$. At the cut frequency $\omega_0 = 1/RC$ the phase shift is $\pm\pi/2$, i.e. the output signal is in quadrature with respect to the input signal. The phase offset in case a) decreases with ω from $+\pi$ to zero and increases in case b) from zero to $-\pi$.

10.4. Double shifter oscillator

In Figure 10.7 OA1 is an inverting amplifier: $|A(j\omega)| = 1$. The other two OA may be seen as the feedback network made by two phase cascaded shifters.

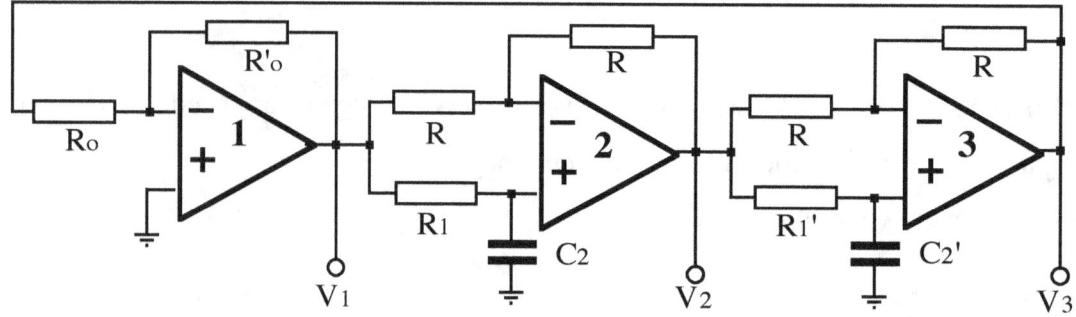

Figure 10.7

With proper choice of the CR dividers the two shifters may be set for a total phase offset of π at some frequency ω_0 producing self-oscillation.

The particular choice $R_1' = R_1$, and $C_2' = C_2$, gives for V_2 a quadrature output, ad for V_3 an output in phase opposition with respect to V_1. The automatic gain control may be achieved by a double diode in parallel to R_o'.

10.5. Quadrature shifter

A phase shifter that provides a constant phase shift of $\pi/2$ for any frequency is shown in Figure 10.8.

The voltage $V_1 = V_{in}Z_c/(Z_c+R_2)$ is amplified with gain $G=1+Z_c/R_1$, so that the transfer function is:

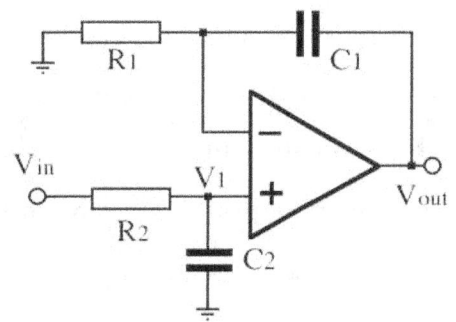

Figure 10.8

$$T(s) = \frac{1}{sR_1C_1} \frac{1+sR_1C_1}{1+sR_2C_2},$$

that, for $R_1C_1 = R_2C_2 = RC$, becomes $T(s)=1/(sRC)$. This circuit is a non-inverting integrator giving an output shifted by $\phi=\pi/2$ with respect to the input. For a constant input amplitude, the output decreases linearly with frequency: $|T(j\omega)|=1/(\omega RC)$.

10.6. Double integrator oscillator

Adding to the previous circuit (OA2) an inverting integrator (OA1) as in Figure 10.9, we obtain a *quadrature oscillator*. The circuit may be seen as figure 10.1, where each one of the two AO may be either the amplifier (A) or the feedback network (β).

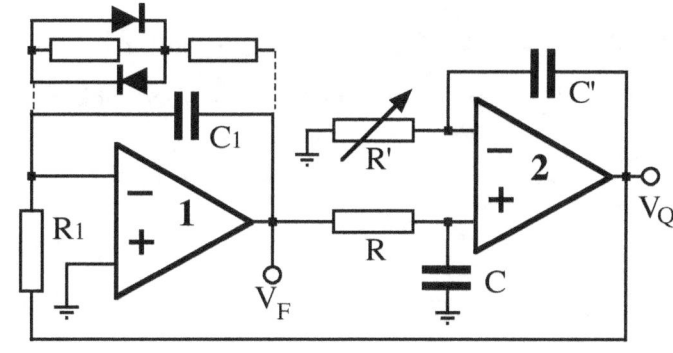

Figure 10.9

Letting $R' = R$ and $C' = C$, we get

$A\beta = 1/(\omega^2 R_1C_1RC)$, which gives an oscillation frequency $\omega_0 = 1/\sqrt{R_1C_1RC}$. Note that the two outputs V_F and V_Q are in quadrature.

The automatic gain control may be achieved placing a double diode in parallel to C_1 and the amplitude may be adjusted by the potentiometer R'.

10.7. Phase shifter oscillator

Figure 10.10 shows the oscillator known as *phase shifter oscillator*. It is made an active differentiator [43] $V_o = A(s)V_1 = -sR_oC\,V_1$ whose negative feedback $V_1 = \beta(s)V_o$ is a double high-pass filter made by two passive differentiators (we neglect here the diodes and R_f that set the automatic gain control).

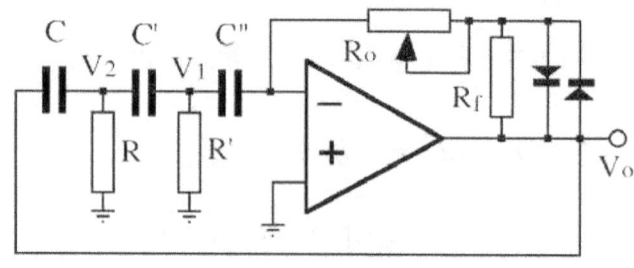

Figure 10.10

The transfer function $\beta(s)$ is the product of the two transfer functions $T_1(s)=V_2/V_o$ and $T_2(s)=V_1/V_2$. Because V_1 is the output of the divider (C',Z_2), (where Z_2 is the parallel of [44] R' and C'') we get $T_2(s)= sR'C'/[1+sR'(C'+C'')]$. Choosing $C=C'=C''$ and $R=R'$, we have more simply $T_2(s)= sRC/(1+2sRC)$.

Because V_1 is the output of the divider (C, Z_1), (where Z_1 is the parallel of R with C' in series to Z_2) we obtain $T_1 = sRC(1+2sRC)/(1+4sRC+3s^2R^2C^2)$.

The condition $|A\beta| =|A(s)T_1(s)T_2(s)|=1$ may be written: $1+4sRC+3s^2R^2C^2 = -s^2R^2R_oC^3$ or $R_o = [4\omega RC-j(1-3/\omega^2R^2C^2)]/\omega^3R^2C^3$; the left side of the last equation is a real number: therefore the imaginary of the right side part must be zero: i.e. $\omega^2R^2C^2=1/3$.

This transforms the condition $|A\beta|=1$ into $R_o = 12R$, and oscillation frequency $\omega_0 = 1/(RC\sqrt{3})$. One should choose R_o slightly larger than 12 R to start oscillation: the two diodes shown in Figure 10.10, and proper trimming of the OA feedback resistance will adjust the oscillation amplitude.

---

[43] See § 8.2.
[44] Note that non-inverting input is a virtual ground.

10.8. Square/triangular wave generator

One comparator and one integrator in a closed-loop as in Figure 10.11 give a generator of triangular wave and of square wave.

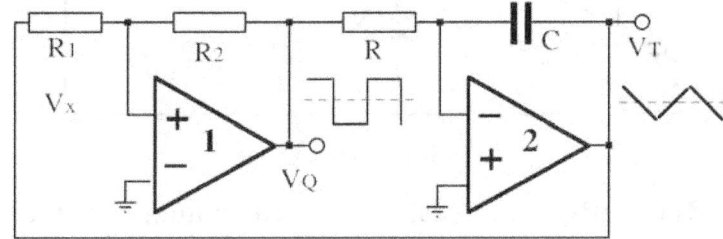

Figure 10.11

The OA1 comparator toggles when the voltage V_x crosses the zero-voltage threshold value set as reference at the inverting input, and the OA2 integrator transforms the constant V_Q output into a ramp. Let us follow the time evolution of the voltages V_Q and V_T, starting from $V_Q = +V_{cc}$: the output V_T of the inverting integrator decreases linearly with time: $V_T(t) = V_T(0) - V_{cc}t/RC$. The voltage V_x is the superposition of sources V_T and V_Q: $V_x = V_T R_2/(R_1+R_2) + V_Q R_1/(R_1+R_2)$, and the comparator toggles when $V_x = 0$, i.e. for $V_T = -V_Q R_1/R_2$, (note that we must set $R_2 > R_1$ to avoid saturation of V_T). This gives the starting value $V_T(0)$ for the *positive* ramp of $V_T(t)$ (because now $V_Q = -V_{cc}$): $V_T(t) = -V_{cc}R_1/R_2 + V_{cc}t/RC$. The next comparator toggling occurs for $V_T(T/2) = V_{cc}R_1/R_2$, at the time $t=T/2$ (the half-period of the square-wave): $V_{cc}R_1/R_2 = -V_{cc}R_1/R_2 + V_{cc}T/2RC$, or $2R_1/R_2 = T/2RC$, that gives $T = 4RC(R_1/R_2)$.

The triangular wave amplitude is $2V_{cc}R_1/R_2$, with frequency $f = 1/T = (R_2/R_1)/4RC$.

The circuit of Figure 10.11 has two drawback: the OA1 input offset voltage V_{os1} gives an offset to the triangular signal, and the OA2 input offset voltage V_{os2} makes not symmetrical the square-wave.

An improved version of this circuit is shown in Figure 10.12, where offset adjustment, amplitude stabilization and symmetry control have been included. The frequency is set by the potentiometer R_F, the amplitude by the potentiometer R_G. The potentiometer R_T corrects the V_T offset and R_Q the square-wave symmetry. Frequency increases by decreasing R_F and the amplitude V_T increases by decreasing R_G.

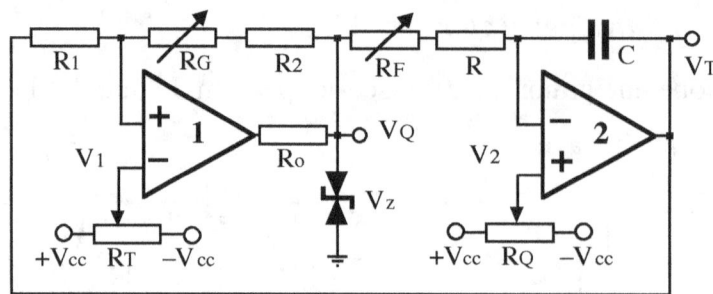

Figure 10.12

Letting $K=(R_2+R_G)/(R_1+R_2+R_G)$ the peak-to-peak amplitude of the triangular wave is $V_{Tpp} = 2V_z(1/K-1)$, with mean value V_1/K, where V_1 is set by adjusting the potentiometer R_T. The frequency is $f = [1-(V_2/V_z)^2]/4[(1/K-1)(R+R_F)C]$, where V_1 is set by adjusting the potentiometer R_Q, so that, for $V_1=V_2=0$ and $R_G=R_F=0$ we get $f=(R_2/R_1)/4RC$, as above.

An equivalent circuit is drawn in Figure 10.13 with an inverting comparator (OA1), with hysteresis and reference voltage $V_R = V_Q R_2/(R_1+R_2)$, plus a non-inverting integrator (OA2).

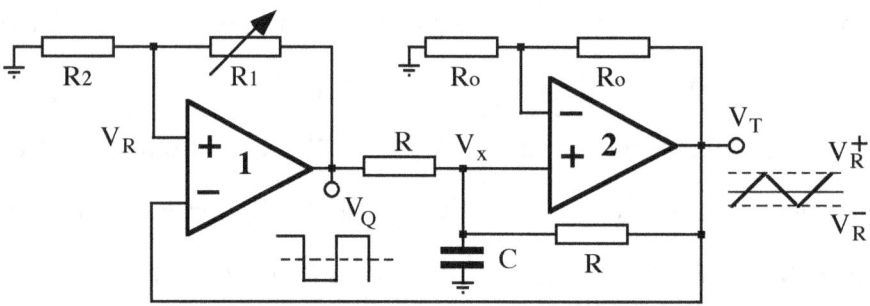

Figure 10.13

The superposition principle gives $V_x = V_{xQ} + V_{xT}$ with: $V_{xQ} = V_Q(R\|Z_c)/(R+R\|Z_c)$, and $V_{xT} = V_T R/(R+R\|Z_c)$, i.e.: $V_{xQ} = V_Q sRC/(2+sRC)$ and $V_{xT} = V_T(1+sRC)/(2+sRC)$

The integrator OA2 amplifies the voltage V_x with gain $G=2$, so we obtain $V_T = 2(V_{xT}+V_{xQ}) = 2[V_T(1+sRC)+V_Q sRC]/(2+sRC)$, that gives the transfer function of AO2: $V_T/V_Q = 2/sRC$, predicting the time evolution of V_T: $V_T(t) = V_T(0) + (2/RC)\int V_Q dt$.

Let us assume t=0 when the comparator switches from $-V_{cc}$ to $+V_{cc}$: at this time the threshold voltage is $V_R^- = -V_{cc}R_2/(R_1+R_2) = V_T(0)$. The voltage V_T start increasing linearly with the law: $V_T(t) = V_{cc}[2t/RC - R_2/(R_1+R_2)]$.

The next comparator toggling occurs after an half-period T/2, when $V_T(t)$ reaches the positive threshold: $V_R^+ = +V_{cc}R_2/(R_1+R_2) = V_T(T/2)$.

The period is therefore $T = 2RCR_2/(R_1+R_2)$, or $T = RC$ for $R_1 = R_2$.

10.9. Quadrature square/triangular wave generator

By cascading two stages of the previous circuit (comparator+ integrator), as in Figure 10.14, we get two square-waves in quadrature and two triangular-waves in quadrature.

The signal V_{Q2} has T/4 delay with respect to V_{Q1}, and V_{T2} delays T/4 with respect to V_{T1}. For $T_1=R_1C_1 \neq T_2=R_2C_2$, the amplitude of the two triangular waves is different and the square-wave is no more symmetric; e.g. for $T_2>T_1$ we have $V_{T1}>V_{T2}$.

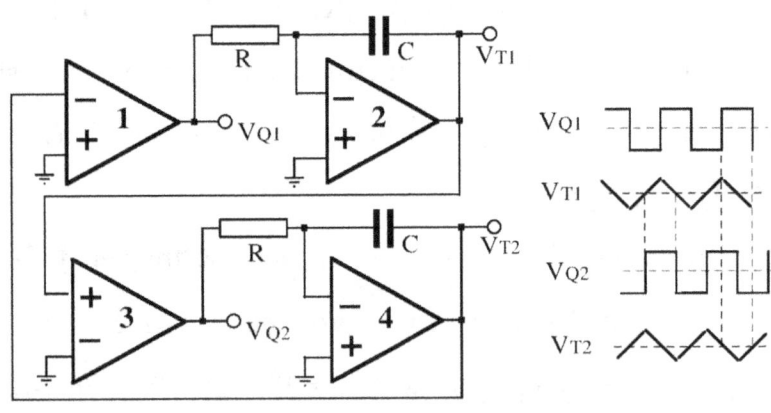

Figure 10.14

10.10. Voltage to frequency converter

Frequency may be modulated by a voltage using a voltage-to-frequency converter as that shown in Figure 10.15. Here the output signal V_3 is made of pulses repeating at the frequency f, proportional to the input voltage V_i.

The circuit is made by an inverting integrator (OA1) and by a non-inverting comparator (OA2) with hysteresis and zero reference voltage V_R (see § 9.2).

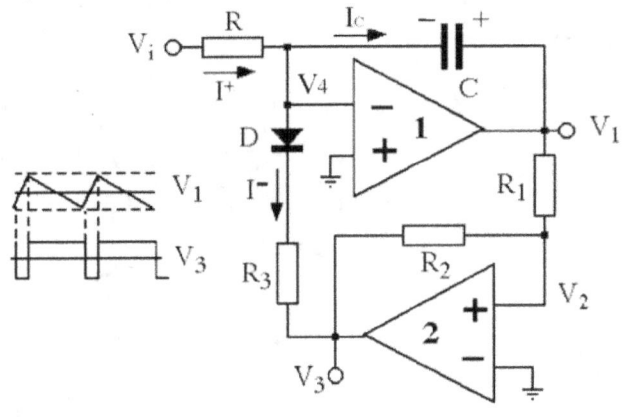

Figure 10.15

Let be t=0 the time at which V_3 switches from $-V_{cc}$ to $+V_{cc}$. Because $V_R=0$, it toggles when its input voltage $V_2(t)$ reaches the *positive* threshold $V_{cc}(R_1/R_2)$; the diode is reverse biased (V_4 is a virtual ground), and OA1 integrates the current $I_C = I^+ = V_i/R$, that gives at the output :

$$V_1(t) = V_1(0) - t(V_i/R)/C = V_{cc}(R_1/R_2) - t(V_i/R)/C \qquad [10.2]$$

The comparator output switches back to $-V_{cc}$ after the time T_1, when $V_1(t)$ reaches the *negative* threshold $V_1(T_1)=-V_{cc}R_1/R_2$.

The equation [10.2] becomes: $-V_{cc}R_1/R_2= V_{cc}(R_1/R_2) - T_1(V_i/R)/C$, that yields the solution $T_1 = 2V_{cc}RC(R_1/R_2)/V_i$.

Let now be t=0 the time at which V_3 switches from $+V_{cc}$ to $-V_{cc}$.

We have $V_1(0) = -V_{cc}R_1/R_2$ and because the diode is forward biased OA1 integrates the current $I_C = I^+ - I^- = V_i/R - V_{cc}/R_3$, giving at the output $V_1(t) = -V_{cc}(R_1/R_2) - t(V_i/R - V_{cc}/R_3)/C$

If we choose $R \gg R_3$ we may neglect $I^+ = V_i/R$, writing:

$$V_1(t) \approx -V_{cc}(R_1/R_2) + t(V_{cc}/R_3)/C \qquad [10.3]$$

The comparator input $V_1(t)$ will cross again the *positive* threshold $V_{cc}(R_1/R_2)$ at the time T_2. The equation [10.3] becomes: $V_{cc}R_1/R_2 \approx -V_{cc}(R_1/R_2) - T_2(V_{cc}/R)/C$, that yields $T_2 \approx 2R_3CR_1/R_2 \ll T_1$.

The signal period is $T = T_1 + T_2 \approx T_1$, and the frequency $f = 1/T \approx V_iR_2/(2V_{cc}RCR_1)$, which is proportional to V_i. A better approximation which takes into account $I_C = I^+ - I^-$ gives: $f \approx k_1(V_i/V_{cc}) + k_2(V_i/V_{cc})^2$, with $k_1 = 2R_2/(2RCR_1)$ and $k_2/k_1 = R_3/R \ll 1$.

10.11. Frequency-to-voltage converter

The inverse process, i.e. the frequency-to-voltage conversion, may be implemented by the circuit shown in figure 10.16, a basic frequency meter for generic a.c. signals with zero mean value $V_i^*(t)$.

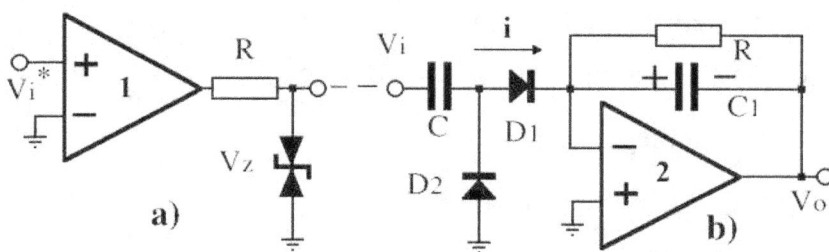

Figure 10.16

In figure 10.16a the OA1 is a zero-reference comparator, with a twin zener load, that transforms the input signal $V_i^*(t)$ into a squared signal $V_i(t)$ of constant amplitude $2V_z$ peak to peak.

At each rising edge of the squared input signal $V_i(t)$ a charge $q = 2V_zC$ is transferred from the capacitor C through the diode D1 into the capacitor C_1 and the same charge is restored into C through the diode D2 at each falling edge of $V_i(t)$, because D1 will be reverse-biased.

This charge transfer corresponds to an *average* current $\langle i \rangle = q/T = CV_i f$, where $T = 1/f$ is the average period of the signal $V_i(t)$ with *average* frequency f.

The capacitor C_1 discharges through the resistor R, during the time between falling and rising edge, but if $RC_1 \gg T$, the output voltage V_o is well approximated by: $V_o = -RCV_i f$. To have a frequency meter with positive output we simply revere the polarity of both D1 and D2 diodes.

11. Phase sensitive detector (lock-in)

The lock-in amplifier is a device that is frequently used to extract weak signals from background noise. Noise sources may be electromagnetic fields due to line power supply or radio-frequency broadcasting, but also acoustical pick-up, thermal noise, shot noise or flicker noise[45].

The line-noise, due to poor shielding or to ground-loops, has Fourier-components at the line-frequency (50Hz or 60 Hz, and multiples). The thermal noise (also named Johnson noise), depends on the source resistance R, on temperature and on the band-width B: its root-mean-square voltage amplitude at room temperature is $V_{RMS}=\sqrt{4RK_BTB} \approx 10^{-4}\sqrt{RB}(\mu V)$. The shot noise, due to the quantum nature of electric charge, depends on the current I and on the band-width B; its root-mean-square current amplitude is $I_{RMS}=\sqrt{2qIB} \approx 10^{-4}\sqrt{IB}(\mu A)$. The flicker noise (also named 1/f noise) decreases with frequency so that it is practically negligible above few tenths of Hz.

We may filter the noise by using narrow band-pass filters tuned at the signal frequency ω_0. The higher is the filter quality factor $Q=\omega_0/(\omega_2-\omega_1)$ the more selective is the filter; however the maximum value for $Q\approx 100$ is limited by instability problems: a slight drift of the central filter frequency (due to temperature changes or aging of components) produces in fact strong signal damping.

An alternative solution is to lock the filter central frequency to the signal frequency: this is the lock-in amplifier technique. A lock-in amplifier needs a *reference signal* V_R that is *synchronous* with the signal to be detected V_S; such signal may be found more easily than it could appear at first sight: quite often in fact the weak signal to be extracted from background noise is produced as *response* to an *excitation* signal that will be available as reference signal. In case of d.c. signals one may always modulate[46] them by "chopping".

The lock-in output is not sinusoidal signal (as for tuned band-pass filters output) but a d.c. voltage whose value is proportional to the amplitude of the detected input signal.

The main advantage of the lock-in is the very high Q-values (of the order of 10^5) even at very low frequencies, where traditional tuned band-pass filters become very expensive.

---

[45] For a nice brief description of electric noise see: *Electronics for the Physicist,* C.G.Delaney, chapt 11. We here only recall that thermal noise is due to the brownian motion of electrons, shot noise is due to the statistical fluctuations of the number of discrete charges flowing in a time unit, while flicker noise may be produced by various different processes.

[46] Choppers are frequently used for example in optical benches where the d.c. light beam crosses a rotating disk with holes, that acts as a on/off switcher at a given frequency: a photodetector sensing part of the beam emerging from the perforated disk provides the reference signal.

11.1. Lock-in with synchronous switch

Let us consider a sinusoidal signal $V_S(t)=V_{SM}\sin(\omega_0 t)$ with angular frequency ω_0, and amplitude V_{SM}, which is *buried* in a background noise V_N with a broad frequency spectrum. The noisy signal may be seen as the superposition V_S+V_N of the signal V_S and the noise V_N.

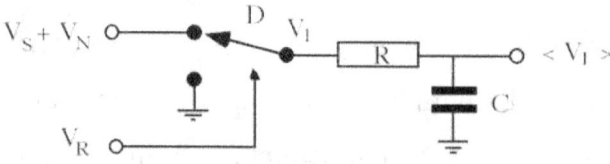

Figure 11.1

Figure 11.1 shows the basic drawing of a lock-in made of a synchronous switch and a low-pass filter: the signal to be processed V_S+V_N is chopped by a *voltage-controlled switch D* and fed to a low-pass RC filter. The switch is controlled by the reference signal V_R *synchronous* with V_S, so that it is passing the signal during the positive half-wave of V_S and it shorts to ground the filter input during the negative half-wave of V_S. This is substantially an *half-wave chopper*.

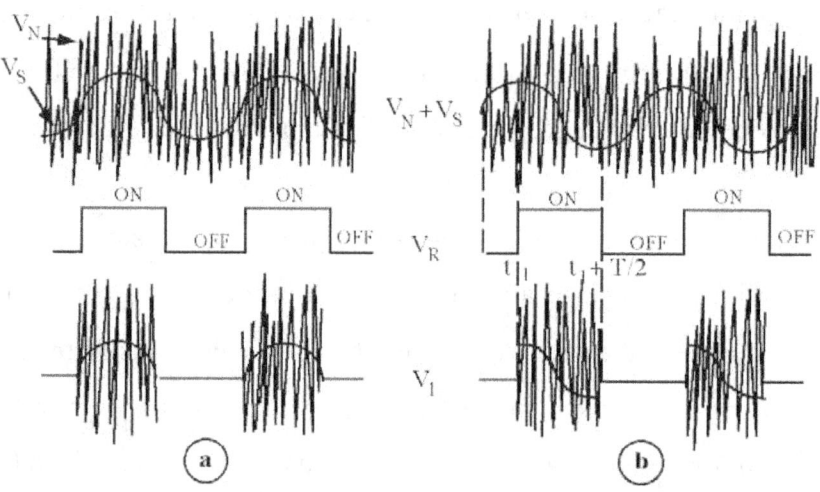

Figure 11.2

The signal shape V_S+V_N (before the switch) and V_1 (after the switch) is sketched in figure 11.2a, where is shown also the waveform of V_S, that in real case is hidden by the noise. After the low-pass filter the mean value is $<V_1> = V_{SM}/\pi$ because the mean value of V_N is zero, if we make the reasonable assumption that the noise has no component synchronous with V_S.

If we set a phase lag between V_R and V_S, i.e. the switch is triggered with a delay t_1, (or a phase shift $\Phi=\omega_0 t_1$) with respect to V_S, the output voltage $<V_1>$ depends, not only on V_{SM}, but also on Φ. An example is shown in figure 11.2b, and an analytic expression of the output is:

$$<V_1> = \frac{1}{T}\int_{t_1}^{t_1+T/2} V_{SM}\sin\omega_0 t \, dt = \left(\frac{V_{SM}}{T}\right)\left[\frac{-\cos\omega_0 t}{\omega_0}\right]_{t_1}^{t_1+T/2} = \left(\frac{V_{SM}}{\pi}\right)\cos\phi \qquad [11.1]$$

Relation [11.1] shows that the lock-in output, at constant V_{SM}, measures Φ, which explains the name of "*phase sensitive detector*" for the lock-in amplifier.

11.2. Lock-in with multiplier

A different lock-in structure is shown in figure 11.3. Here the block marked by $\times$, replacing the synchronous switch of figure 11.1, is a *multiplier*, i.e. a device that gives an output voltage $V_1(t)$ proportional to the product of the input voltages $V_S(t)$ and $V_R(t)$: $V_1(t) = k\, V_S(t) \times V_R(t)$.

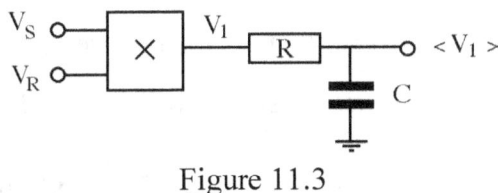

Figure 11.3

Frequently, in the commercial IC multipliers, the value of the factor k is 1/10, but here we'll assume $k=1$, for simplicity.

When $V_S(t)$ and $V_R(t)$ are sinusoidal functions: $V_S(t) = V_{SM}\sin\omega_S t$ and $V_R(t) = V_{RM}\sin\omega_R t$, we get:

$$V_1(t) = V_{SM}V_{RM} \sin\omega_S t\, \sin\omega_R t = V_{SM}V_{RM}[\cos(\omega_S - \omega_R)t - \cos(\omega_S + \omega_R)t]/2 \qquad [11.2]$$

where we used the Werner trigonometric formulas to compute the $\sin\omega_S t\, \sin\omega_R t$ product.

The output signal V_1, has two components, with frequencies that are the sum and the difference, respectively, of the two frequencies of input signals.

In the particular case $\omega_S = \omega_R = \omega_0$, with a phase shift Φ between input signals, we get:
$V_1(t) = V_{SM}V_{RM}[\cos\Phi - \cos(2\omega_0 t + \Phi)]/2$.

Here the output has a d.c. component ("zero-frequency term") that depends on the phase shift, and a component that is the second harmonic of the signal frequency ω_0.

At the low-pass output (under the condition $RC \gg 1/2\omega_0$) we get

$$\langle V_1 \rangle = (V_{SM}V_{RM}/2)\cos\Phi. \qquad [11.3]$$

Relation [11.3] gives the same dependence on Φ as relation [11.1], but here the lock-in output depends also on the amplitude V_{RM} of the reference signal. A reliable measurement of the detected signal amplitude therefore requires not only a *stable phase shift* but also a *stable amplitude for the reference signal*.

The transfer function of this lock-in has the spectrum shown in figure 11.4. The bandwidth $\Delta\omega$ of the band-pass filter, centered at ω_0, is determined by the time constant RC of the low-pass filter.

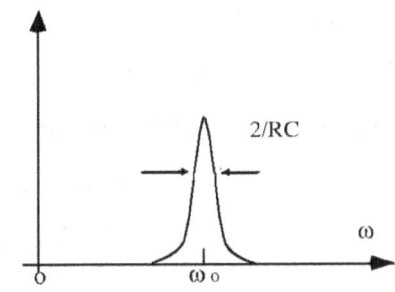

Figure 11.4

This means that the noise components with frequencies ω_i, with $|\omega_i-\omega_o|<1/RC$, modulate the output voltage $<V_1$. One might see these components with frequencies very close to ω_o, as *quasi-synchronous components* equivalent to a synchronous signal with phase shift slowly changing with time.

We may analyze again the circuit of figure 11.1 assuming the switch to be a multiplier with a reference signal that assumes values 0 and 1 (a square-wave with mean value $<V_R>=1/2$).

A generic periodic signal, with period $T=2\pi/\omega_R$ may be written in terms of Fourier components:

$$V(t) = a_0 + \sum_{n=1}^{\infty} a_n \sin(n\omega_R t + \Phi_n) \; , \qquad [11.4]$$

where a_0 is the mean value and a_n are the Fourier amplitudes.

In our case $a_0=1/2$, the even amplitudes are zero and the odd amplitudes are $a_n=2/\pi n$. Therefore V_R may be written:

$$V_R(t) = \tfrac{1}{2} + \tfrac{2}{\pi}\left[\sin\omega_R t + \tfrac{1}{3}\sin 3\omega_R t + \tfrac{1}{5}\sin 5\omega_R t + ...\right], \qquad [11.5]$$

and the output signal $V_1(t = V_S(t) \cdot V_R(t)$ becomes:

$$V_1(t) = \tfrac{1}{2}V_{SM}\sin\omega_S t + \tfrac{2}{\pi}V_{SM}[\sin\omega_S t \sin\omega_R t + \sin\omega_S t \sin 3\omega_R t/3 + ...]. \qquad [11.6]$$

If the reference signal V_R is synchronous with V_S, i.e. $\omega_R=\omega_S=\omega_o$, in relation [11.6] survives a single d.c. term, and for $RC>>1/\omega_o$ we obtain at the filter output for $<V_1>$ again relation [11.1]. We note that if the noise V_N includes a d.c. term V_{OS}, i.e. $V_N + V_S = V_{OS} + V_N(t) + V_{SM}\sin\omega_o t$, then the offset will appear also at the lock-in output:

$$<V_1> = \tfrac{1}{2}V_{OS} + \tfrac{1}{\pi}V_{SM}\cos\Phi. \qquad [11.7]$$

Relation [11.6] shows that all the odd harmonics $(2n-1)\omega_o$ of V_S contribute to $V_1(t)$, so that the transfer function of this lock-in has the spectrum shown in figure 11.5.

The bandwidth $\Delta\omega$ of the peaks centered at $0, \omega_o, 3\omega_o, 5\omega_o,...$ is determined by the time constant RC of the low-pass filter: $\Delta\omega=2/RC$. This means that the noise components with frequencies ω_i, with $|\omega_i-(2n-1)\omega_o|<1/RC$, modulate the output voltage $<V_1$.

The lock-in with (0,1) multiplier may be seen as a parallel of infinite numbers of lock-ins with sinusoidal multiplier and with reference signals made by odd harmonics of the signal to be detected.

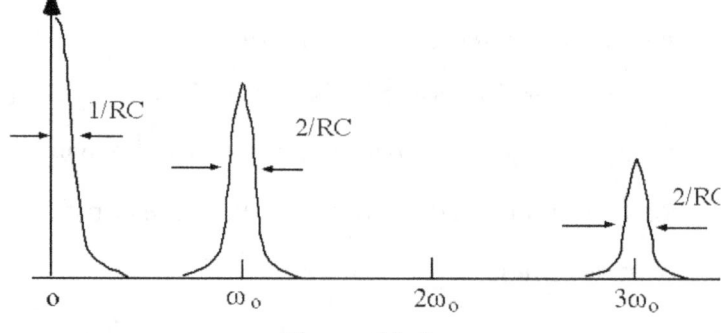

Figure 11.5

The (0,1) square wave reference method introduces more noise band-pass windows with respect to the sinusoidal reference method. However it is easier to stabilize the amplitude of a square wave than the amplitude of a sinusoid.

The value of the time constant RC is limited only by the required response time τ ($\tau \approx 5\,RC$), so that choosing $RC \gg 1/\omega_0$ we may obtain very high Q values ($\omega_0/\Delta\omega = 2RC\omega_0$ up to 10^5).

11.3. Lock-in with multiplier ±1

Using as reference signal a symmetric square wave (e.g. ±1) we may get a further improvement.

Figure 11.6 shows a modification of figure 11.1, where two amplifiers followed by a voltage controller switch behave as a multiplier by ± 1.

The analysis of the behavior of this circuit is the same as that made for circuit of figure 11.1, with the difference that in the Fourier series [11.4] we have now $a_0=0$ and $a_n = 4/\pi n$, so that the mean value of the product $V_1(t) = V_S(t)\,V_R(t)$ becomes (even in presence of an offset V_{OS} in the input signal):

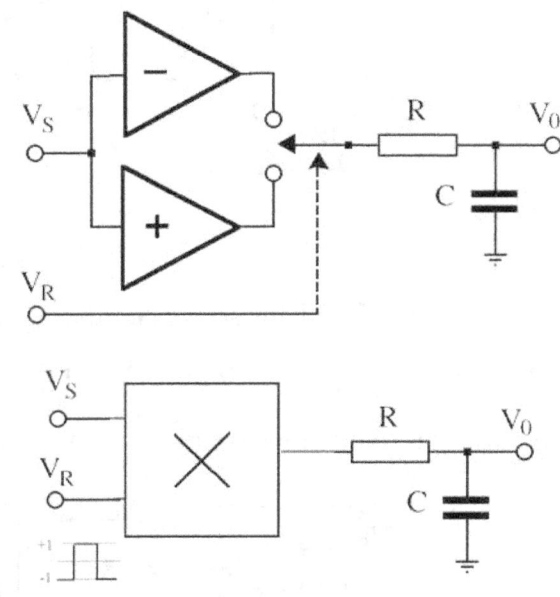

Figure 11.6

$$< V_1 > = \frac{2}{\pi} V_{SM} \cos\Phi \,. \qquad [11.8]$$

The transfer function of this circuit has the spectrum depicted in figure 11.7, where the peak at zero frequency disappears, allowing much better rejection of offset and flicker noise from V_S.

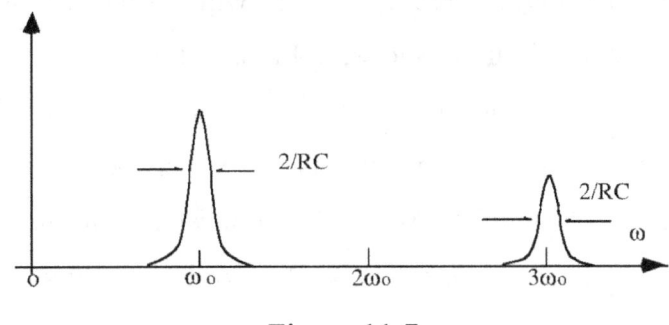

Figure 11.7

In Figure 11.6 the voltage-controlled switch may be implemented by a relay, or a pair of FET or CMOS Analog Switches [47]. A simpler version of circuit 11.6 is shown in Figure 11.8, where a single OA (see § 4.5) plus a voltage-controlled switch implements the required multiplier (±1 square wave).

---

[47] A brief description of Field Effect Transistors (FET) is given in Appendix A.7; for Analog Switches see §13.3.

When the switch is ON the OA is an inverter (G=−1), when the switch is OFF the OA is a follower (G = +1). The accuracy of this circuit is limited by the non-ideal characteristics of electronic switches ($R_{ON} \neq 0$ and $R_{OFF} \neq \infty$): in case of CMOS the value of R_{ON} is of the order of fractions of kΩ and for R_{OFF} several MΩ. Therefore the resistor R_1 must be selected in order to satisfy the conditions $R_{ON} \ll R_1 \ll R_{OFF}$.

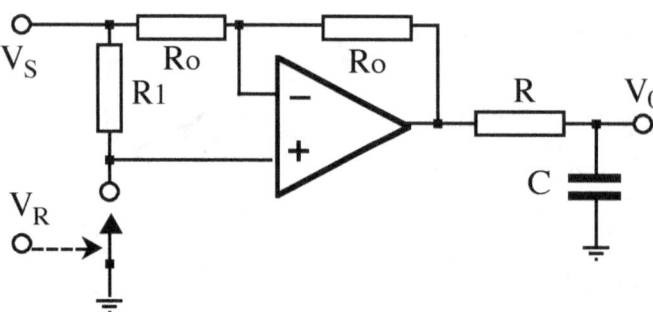

Figure 11.8

Considering Figure 4.7 and Figure 11.8, the gain becomes (accounting for the finite values of R_{ON} and R_{OFF}): $G^- = [2R_{ON}/(R_1+R_{ON}) - 1] \approx -1$ or $G^+ = [2R_{OFF}/(R_1+R_{OFF}) - 1] \approx +1$.

To improve the approximations we may use two analog switches, as shown in Figure 11.9.

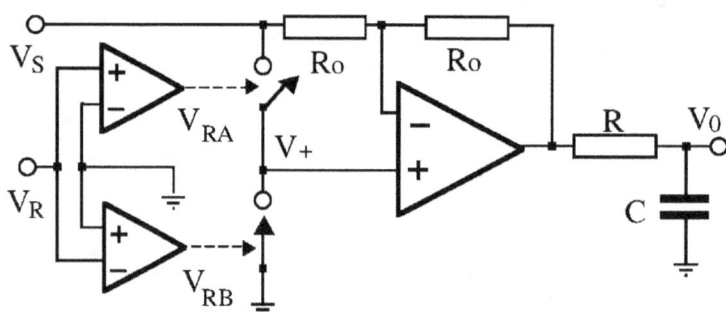

Figure 11.9

Here the two switches are driven in phase-opposition by the two comparators (one inverting and one non-inverting) with reference voltage at ground.

The single channel chopper shown in figure 11.9 may be replaced by a twin-channel chopper followed by a differential (low-pass) amplifier as in the circuit shown in Figure 11.10.

Here the quad analog switch is driven in phase opposition by the two comparators so that the input signal V_S is alternately fed to the differential amplifiers inputs every half-period. This configuration is particularly useful when the source signal V_S is *floating* (not referred to ground voltage) : in this case two wires will feed the differential signal ($V^+_S - V^-_S$) to the analog switch inputs.

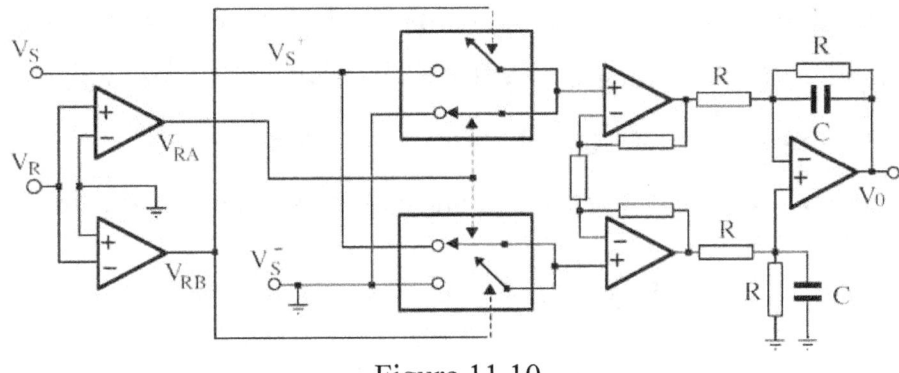

Figure 11.10

Another configuration is shown in Figure 11.11, which is essentially a *full-wave chopper* that duplicates the circuit of Figure 11.1, with a differential amplifier that reads the voltage difference at the outputs of the two RC filters.

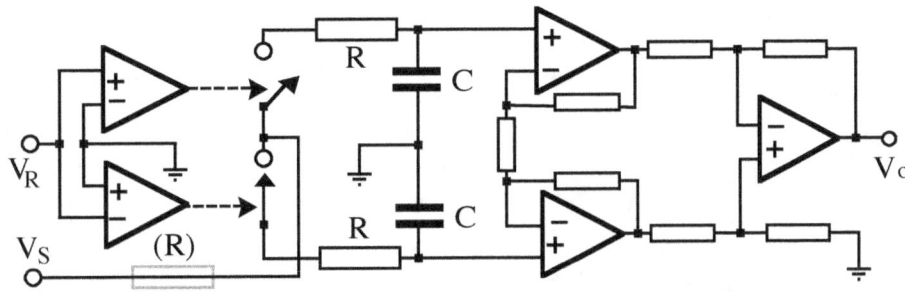

Figure 11.11

The two resistors named R may be replaced by a single resistor (R) if $R_{ON}C \gg \tau_s$, where τ_s is the *switching time* of the two analog switches (to avoid discharging the capacitors during the fraction of τ_s when the two capacitors are shorted by $2R_{ON}$).

11.4. Synchronous filter

Another circuit that may efficiently increase the signal-to-noise ratio is the *synchronous filter* shown in Figure 11.12. This circuit differs substantially from a lock-in: *it gives an output that is a square wave* synchronous with the signal V_S to be detected, and with an amplitude proportional to the V_S amplitude.

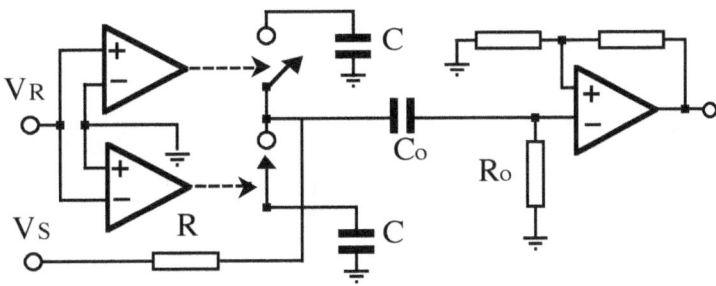

Figure 11.12

If ω_0 is the angular frequency of the V_S, the time constant RC of the two low-pass filters must satisfy the relation $RC \gg 1/\omega_o$, while time constant R_oC_o of the high-pass filter, feeding the output non-inverting amplifier (and deleting eventual offset), must satisfy the relation $(1/R_oC_o) \ll \omega_o$. The output square wave $V_o(\omega_o)$ has a peak-to-peak amplitude equal to $2|V_S|$.

This circuit is often used as a *chopping preamplifier* in sophisticated lock-in circuits

12. Digital electronics: elementary notions

This chapter offers a fast outline of the basic elements that may be found in digital circuits: only a small fraction of the large number of IC devices commercially available will be analyzed.

This brief digest should however be sufficient to give at least an idea of the working principles of most of IC digital devices and to provoke some curiosity into the reader who might deepen his knowledge elsewhere [48].

12.1. Logic circuits

Digital logic circuits are those circuits where only two stable states are possible in any point of the network: e.g. a transistor which is ON (saturated) or OFF (not conducting) or a diode forward or reverse biased,...

Normally we consider voltages, not currents, and we define a state as "high" ("H" or "TRUE", or "1") when the voltage level is above some high threshold value, and we define it as "low" ("L" or "FALSE", or "0") when it is below some low threshold value.

In the logic circuits made with bipolar transistors (TTL =Transistor-Transistor-Logic) [49] that are powered at +5 V high threshold value is about +2.0 V and the low threshold value is about +0.8 V. In the logic circuits made with CMOS FET (Complementary-Metal-Oxide-Semiconductor Field-Effect-Transistor) the threshold voltages depend on the low V_{SS} and high V_{DD} bias voltages. Normally $V_{SS}=0\,V$, and V_{DD} may be any value between +5 V and +15 V: generally we choose $V_{DD}=+5\,V$ or $+12\,V$.

We must distinguish the input threshold values from output stable values: a margin must be provided to warrant proper working in presence of noise, temperature changes, manufacturer's tolerance... This means that the minimum output voltage in the H state of any device must always be higher that the high threshold input value for any device; and the maximum output voltage in the L state of any device must always be lower than the low threshold input value for any device.

A diagram with the limit values for the input/output threshold voltages in TTL and CMOS circuits with $V_{DD}=12\,V$, is shown in Figure 12.1

---

[48] More detailed discussions may be found in *Microelectronics*, by J. Millman and A. Grabel, or in in *TTL Cookbook* or in *CMOS Cookbook*, by D. Lancaster, or in *Digital Electronics* by W.G. Young.

[49] Se for more details http://en.wikipedia.org/wiki/Logic_gate

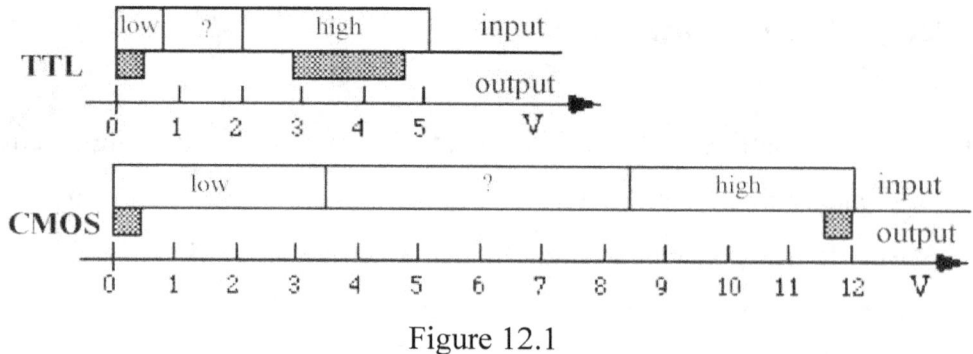

Figure 12.1

When the voltage at input has a value within the range marked with a question-mark the device may detect the input signal either as "high" or "low", so that the state of its output is random.

Any complex digital circuit may be split into basic blocks named *logic gates*. The basic gates are of three kinds : NOT (inverter), AND and OR; the corresponding graphic symbols and behavior (truth tables) are shown in Figure 12.2.

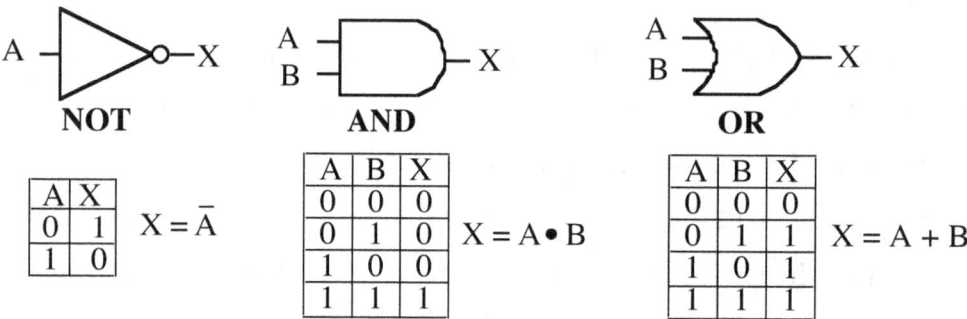

Figure 12.2

The matrices in the lower part of Figure 12.2 are named *truth tables* and they define the behavior of each gate, i.e. the relation between the output logic value X and the given values of inputs A and B. For example: if the output X is the result of "A AND B" (also written as "X = A•B"), this means that X is "high" only when both A and B are "high" at the same time.; if X = A OR B (also written as "X = A+B", this means that X is "high" when A is "high" or when B is "high".

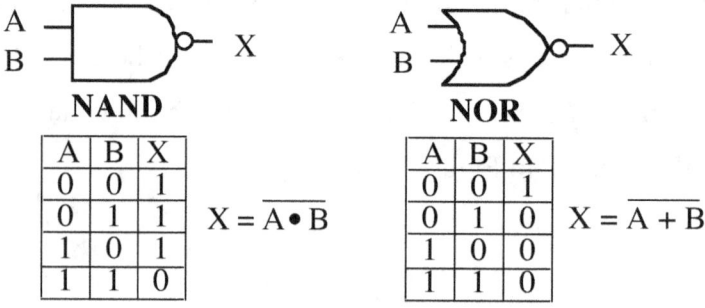

Figure 12.3

A bar placed over a logic variable symbol means its logical *negation* , e.g. if A is "high" , then $\overline{A}$ (=A *negate*) is "low". In the graphic symbols the negation is marked by a small circle at the gate output, which indicates an added NOT gate. For example adding a NOT to the OR output we get a NOR gate, and adding a NOT gate to a AND gate we get a NAND gate, whose truth tables are shown in Figure 12.3.

Using the truth tables we see that a NAND gate may be made by two inverters added at the OR inputs, and a NOR by two inverters added at the AND inputs. These equivalences, shown in Figure 12.4, are the *De Morgan theorems*.

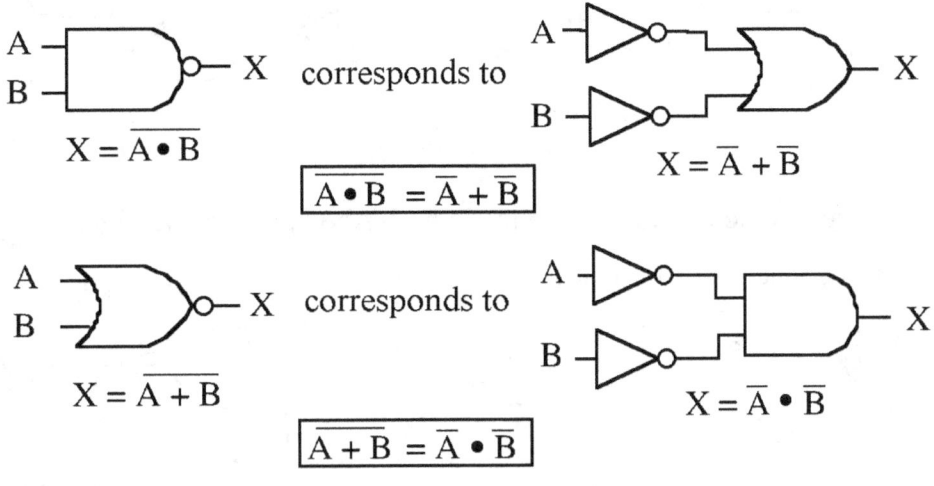

Figure 12.4

Another gate frequently used is the EXCLUSIVE OR, defined by relation $A \oplus B = \overline{A} \cdot B + A \cdot \overline{B}$ and by the truth table in Figure 12.5; it is made either of 2 NOT + AND + OR, or of AND + 2 NOR, or NAND + AND + OR:

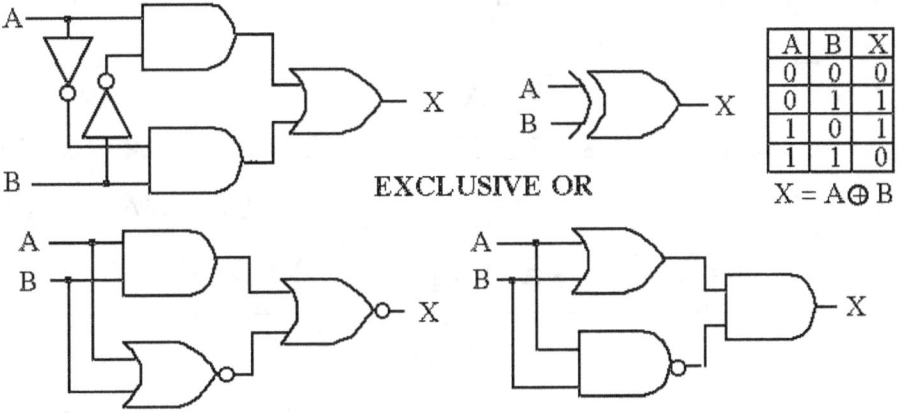

Figure 12.5

A NAND gate with the two inputs shorted, or one input "high" is a NOT gate un inverter. Using two NANDs we may get one AND, with three NANDs we may get one OR, with four NANDs we may get one NOR and with six NANDs we may get one EXCLUSIVE-OR, as shown in Figure 12.6.

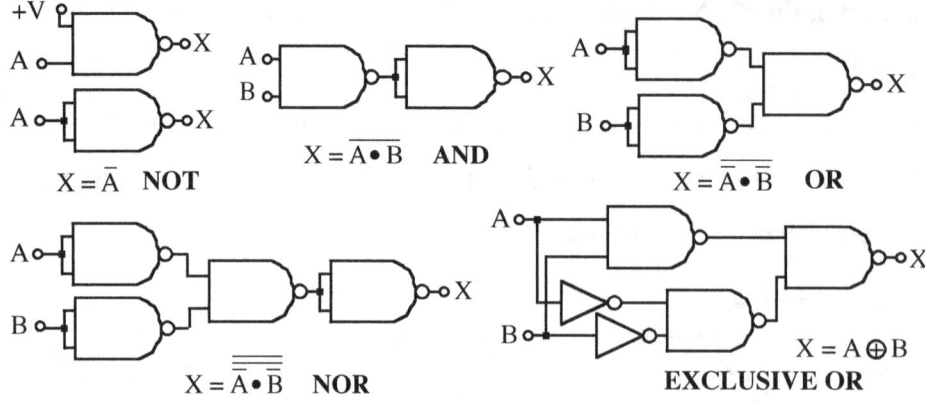

Figure 12.6

This proves that any logic circuit may be made by NAND gates only. Figure 12.7 shows that any circuit may also be made of NOR gates.

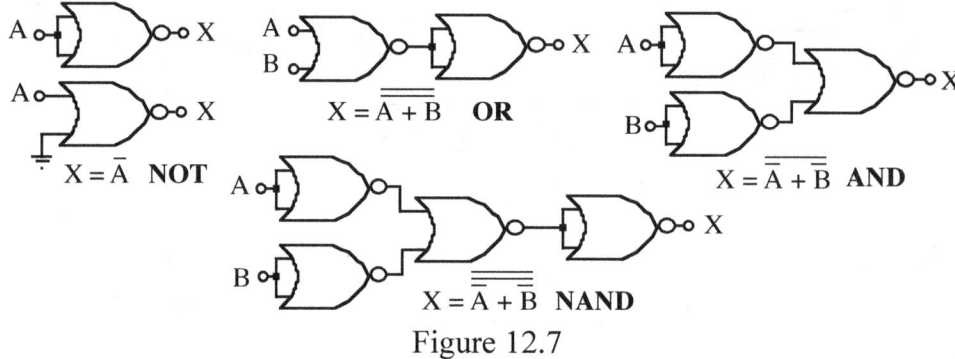

Figure 12.7

The analysis of logic circuits may be made easier [50] using the identities shown in table 12.1.

$A \cdot \overline{A} = 0$	$A + \overline{A} = 1$	$\overline{A + B} = \overline{A} \cdot \overline{B}$	$A \cdot B \cdot C = A \cdot (B \cdot C)$
$A \cdot A = A$	$A + A = A$	$\overline{A \cdot B} = \overline{A} + \overline{B}$	$A + B + C = A + (B + C)$
$A \cdot 1 = A$	$A + 1 = 1$	$A + A \cdot B = A$	$A + B \cdot C = (A + B) \cdot (A + C)$
$A \cdot 0 = 0$	$A + 0 = A$	$A + \overline{A} \cdot B = A + B$	$A \cdot (B + C) = (A \cdot B) + (A \cdot C)$

Table 12.1

Two typical internal structures of a NOT (TLL inverter) are shown in figure 12.8

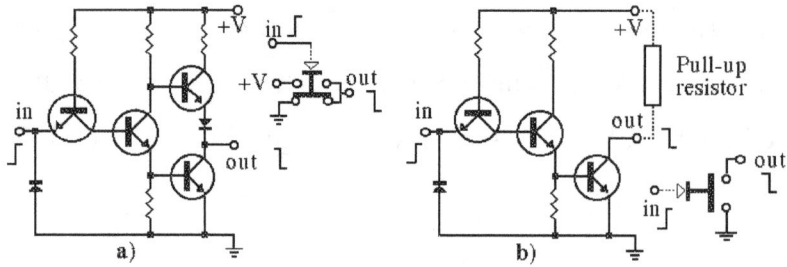

Figure 12.8

---

[50] A compact but complete collection of rules for Boolean algebra may be found in http://www.asic-world.com/digital/boolean1.html#Symbolic_Logic

In Figure12.8a is drawn the standard *Totem-pole* configuration that switches the output between V^+ and ground, and in Figure12.8b the *open collector* configuration, that requires a *pull-up* resistor. The first configuration does not allow connecting more gate-outputs together, while the second one allows to use many inverters with common output (which makes a NOR with many inputs). The drawback is that the pull-up resistor reduces the device speed.

Floating TTL inputs go "high"; the TTL inputs shorted to ground inject a current of about 1.6 mA, and shorted to +V drain a negligible current (≈ 0.04 mA). The power available at one TTL output can drive up to 10 gates (we say it has a *fan-out* of 10), with a maximum current to ground of about 0.4 mA, and a maximum current drained from +5 V of about 16 mA (with output "low"). The main families[51] of TTL gates are 74xx and 54xx where xx stays for the number that specify the device. The family 54xx extends the 74xx working temperature range from (0 °C ÷ +70 °C) to (−55 °C ÷ +125 °C).

A label (L, H, S, LS, F, AS) between 74 / 54 and the number xx, distinguishes sub-families that differ for speed and power: Low-power (L), which traded switching speed (33ns) for a reduction in power consumption (1 mW) . *High-speed* (H), with faster switching than standard TTL (6ns) but significantly higher power dissipation (22 mW). *Schottky* (S), operated more quickly (3ns) but had higher power dissipation (19 mW) *Low-power Schottky* (LS) good combination of speed (9.5ns) and low power consumption (2 mW. *Fast* (F) and *Advanced-Schottky* (AS) speed up the low-to-high transition.

The families CMOS (74HCxx *high speed*, 74HCTxx *high speed TTL compatible*) offer a current output of about 20 mA. CMOS gate inputs do not drain current; input not used should be shorted to ground or to V^+ to avoid possible damage due to static electricity charge. CMOS gate inputs do not drain current; input not used should be shorted to ground or to V^+ to avoid possible damage due to static electricity charge.

name	type	t (ns)	V_{cc} (V)	power (mW)
74xx	TTL-Normal	10	5	10
74Hxx	TTL-High Speed	6	5	22
74Lxx	TTL-Low Power	33	5	1
74Sxx	TTL-Schottky	3	5	19
74LSxx	TTL-Low Power Schottky	10	5	2
74HCTxx	CMOS (TTL input)	10	2-6	.001
74HCxx	CMOS (TTL pin compatible)	10	2-6	.001
40xx	CMOS	100	3-18	<.001

---

[51] An outline of different logic families (RTL, DTL, ECL, TTL, IIL, CMOS, HC, ...) may be found in http://en.wikipedia.org/wiki/Q_factor

A short list of TTL and CMOS logic gates is given in Appendix D [52].

12.2. Bistable circuits: the flip-flop

A *flip-flop* or *latch* is a circuit that has two stable states and can be used to store state information[53]. Two inverters in a closed loop as in Figure 12.9 make a *bistable multivibrator*, also named *RS flip-flop*, where the acronym RS stays for SET-RESET.

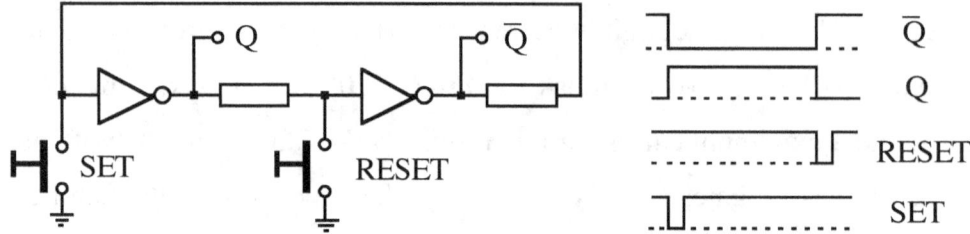

Figure 12.9

The outputs Q and $\overline{Q}$ are stable states that toggle when the corresponding input is grounded by the SET switch or the RESET switch. By grounding SET we get Q = "1", and by grounding RESET we get Q = "0".

This circuit has *memory*, i.e. it toggles when R (or S) is shorted, only if previously S (or R) was shorted. The resistors in Figure 12.9 are needed to protect the inverters in case both switches are shorted to ground, which gives Q = $\overline{Q}$ = "1".

Another RS flip-flop circuit is shown in Figure 12.10, in two different configurations, made with two NAND or two NOR, respectively.

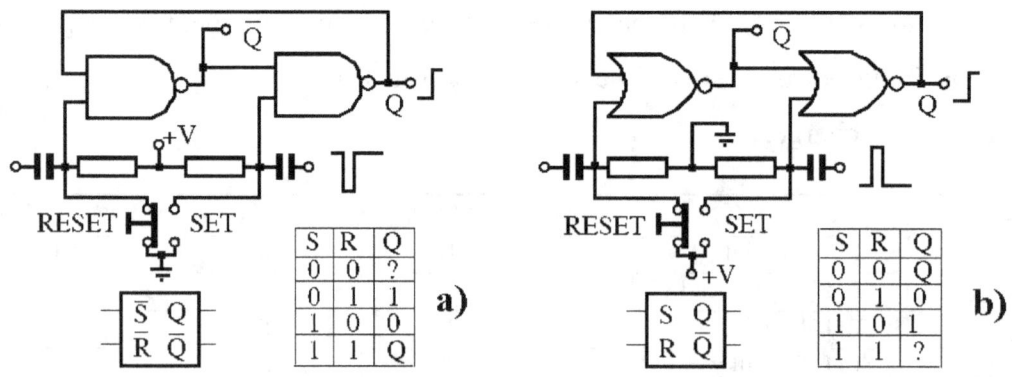

Figure 12.10

Toggling may be triggered by a pulse (negative in the first case and positive in the second case). When the pulse is applied through a coupling capacitor (as in figure 12.10) it is named *edge triggering* instead of *level triggering*. With edge triggering the pulse duration has no effect (e.g. a

---

[52] See http://en.wikipedia.org/wiki/List_of_7400_series_integrated_circuits, and
http://en.wikipedia.org/wiki/4000_series
[53] See also http://en.wikipedia.org/wiki/Flip-flop_%28electronics%29

RESET pulse may be effective even if the SET pulse is not terminated.

The question mark in the third column in truth table defines *disallowed* state (or *forbidden state:* both outputs in the same state: $Q=\overline{Q}$). In fact (from Figure 12.2) the NAND output is 1 when any input is 0, and the NOR output is 0 when any input is 1. The symbol Q in the third column in truth table defines a stable state (either 0 or 1).

12.3. Synchronous flip-flop

The basic synchronous flip-flop is drawn in Figure 12.11a. The name *synchronous* means that the SET (or RESET) command is executed when the CLOCK goes high.

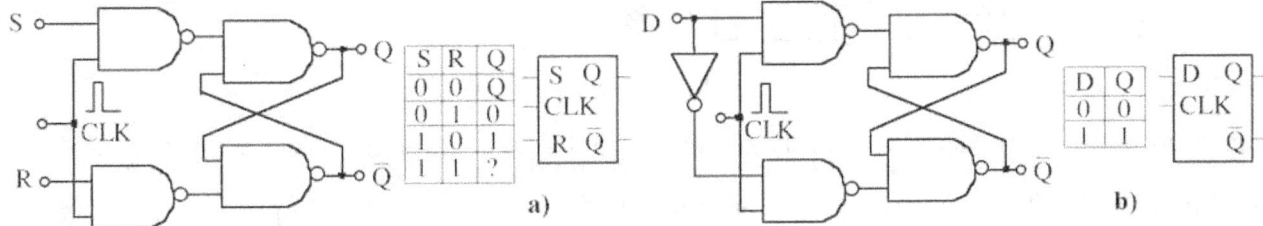

Figure 12.11

With CLOCK enabled the two NAND gates driven by S and R behave as inverters, with CLOCK disabled the output of these NAND gates is "1". The other NAND gates are connected as in Figure 12.10. The disallowed state is for $S=R=$ "1", that gives $Q=\overline{Q}=$"1".

A modification of this circuit, shown in Figure 12.11b, where an inverter connects R to S is the *type-D Flip-Flop*, where D means "data" or "delay" because, the input D value is transferred to the outputs Q, with a delay.

The synchronous latch of Figure 12.11a *allows multiple toggling* during a single CLOCK pulse. A configuration that avoids multiple toggling is the *master-slave* flip-flop shown in Figure 12.12. Here two identical synchronous latches in series are triggered by the CLOCK pulse: an inverter provides the needed counter phase trigger for the two latches.

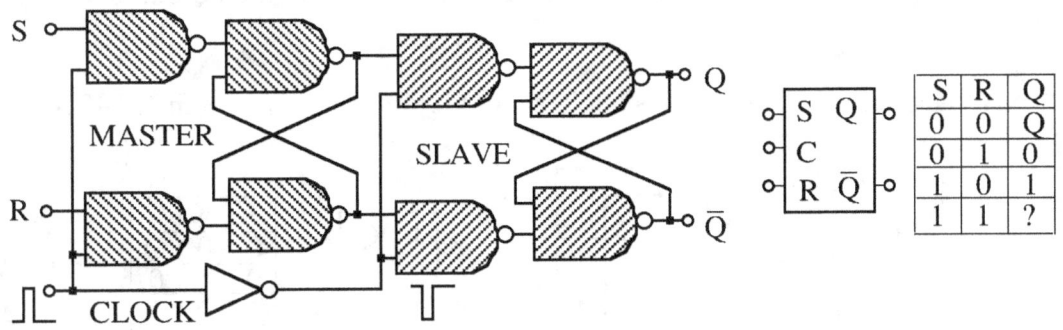

Figure 12.12

The first flip-flop (*master*) acquires the logic values set by R and S during the CLOCK "high" pulse, and its state is transferred to the second flip-flop (*slave*) when the CLOCK goes low. If more than one SET/RESET signal is fed during the clock "high" pulse, only the last state of the input logic values controls the outputs when the clock goes down. The disallowed state is again for R = S = "1", that gives "1" to both outputs of the *master* latch.

If R and S ports are connected by an inverter, as in Figure 12.13, we obtain again a *type D flip-flop*.

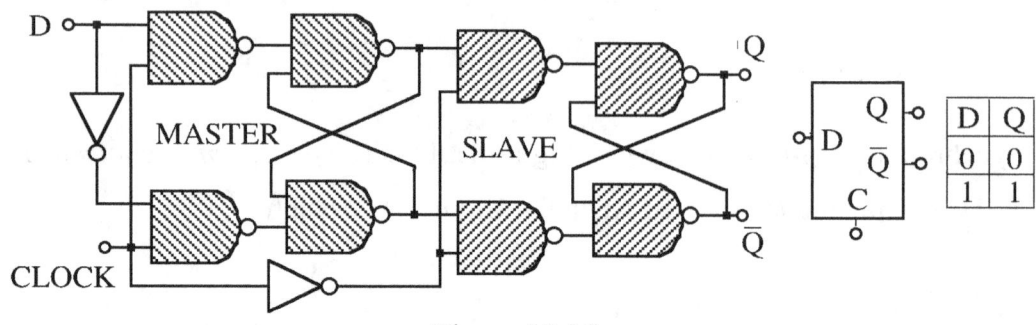

Figure 12.13

The truth table of this circuit is easily obtained from that of the master-slave flip-flop with the condition $D = S = \overline{R}$. The type-D flip-flop transfers the logic value of the input D to the output Q when the clock goes low; it is therefore a *negative-edge triggered* device. Similar circuit obtained by replacing the NAND gates with NOR gates is *positive-edge triggered*.

If the output $\overline{Q}$ is shorted to the input D, as in Figure 12.14a, we get a *divider by two* (also named *Type-T* flip-flop): the input is the CLOCK port and the outputs toggle at each input pulse; the output is always a square wave, for a constant frequency clock. The same device is obtained from an RS Flip-Flop by feeding back the $\overline{Q}$ to S and the Q to R, as in figure 12.14b.

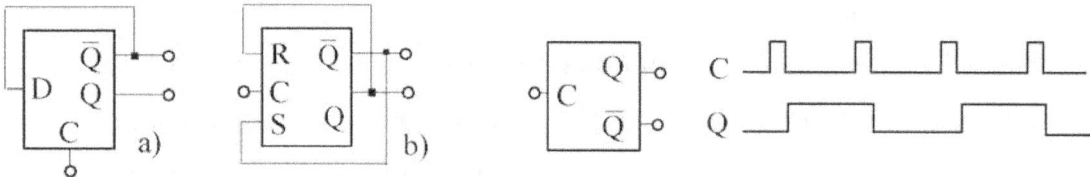

Figure 12.14

Many (n) cascaded dividers by two make a divider by 2^n. Note that the output is always a square wave.

Adding two AND gates to a synchronous latch, as in Figure 12.15, we obtain a J-K flip-flop.

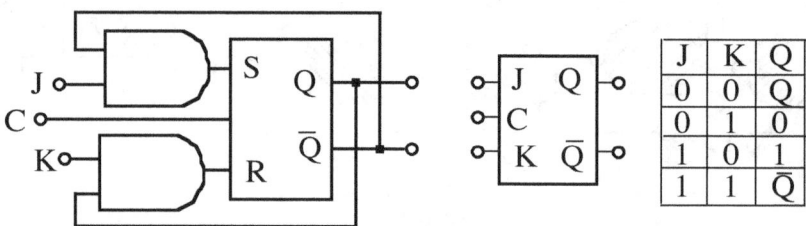

Figure 12.15

If J or K are "1" the AND gates transmit to S and R the logical values of $\overline{Q}$ and Q; if they are "0" they gates transmit to S and R the logical value "0". Therefore S and R never take the logical value "1" at the same time, and the disallowed state is removed.

A J-K flip-flop becomes a divider-by-two when both J and K are "1", and it becomes a type-D latch if J and K are connected by an inverter.

If J and K are shorted we get a type-T flip-flop, (T stays for *toggle*): when the T-port is "1" the output toggles at the CLOCK rising, when the T-port is "0" the toggling is disabled. In Figure 12.16 an example of the clock, toggle and output Q signals are shown.

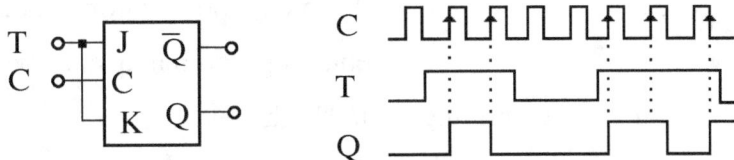

Figure 12.16

In the 74xx family, dual J-K flip-flop: 7473 and 7476, 7474 is a dual D-type flip-flop.

In the CMOS family: 4013 is a dual D-type, 4027 a dual J-K and 4043 is a quad RS latch type NOR, and 4044 is a quad RS latch type NAND (see figure 12.10), 4049 is an hex inverter.

The 7476 and 7474 gates have PRESET and CLEAR inputs that, when are set to low level, force the Q output to high level and to low level, respectively; these inputs are normally kept at high level: if both are low the device is in disallowed state The 7473 device has only CLEAR input. PRESET and CLEAR are implemented also in 4027.

12.4. Monostables

A monostable is a device that gives an output pulse with preset width (*one-shot* pulse) when a suitable signal (*trigger*) is fed to the input. An example made of two NOR gates is shown in Figure 12.17. The trigger is applied to the first gate whose output is fed, through an high-pass filter, to the second gate input.

The trigger pulse is any signal with a fast rising edge with amplitude lager that 3V.

The working principle is the following. The output of gate 2 is "0" because its input B_2 is "1" (due to the bias resistor R). The output of gate 1 is "1" because its input B_1 is "0" (due to the bias resistor R_1). When the trigger pulse

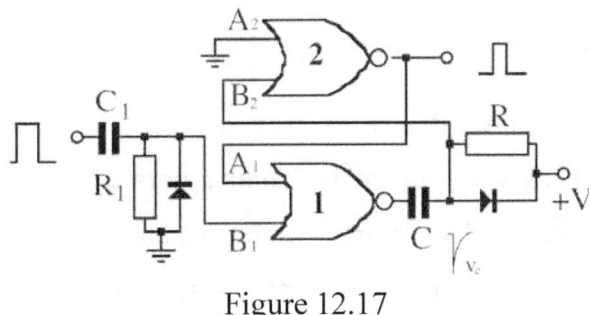

Figure 12.17

toggles to "0" the gate 1 output, and to "1" the gate 2 output, the capacitor C starts to be charged by the resistor R and when the voltage of input B_2 reaches the threshold V_{TH}, the gate 2 output toggles back to "0".

The pulse width is $RC \ln[V/(V-V_{TH})]$, where V is the bias voltage and V_{TH} the threshold voltage. The diodes protect the gates inputs from over voltages due to the high-pass filters.

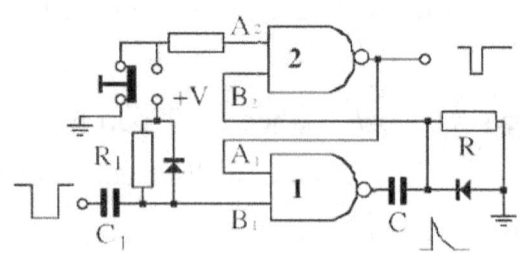

Figure 12.18

A similar one-shot circuit, made with NAND gates, is shown in Figure 12.18, where we added an *enabler* switch at the A_2 input. If A_2 is kept "high" (one-shot enabled) the output toggles at the B_2 spike trigger, if A_2 is kept "low" (one-shot disabled) no toggling occurs. An equivalent enabling command may be set, in the previous circuit with NOR gates, with a switch pulling A_2 to "high" level.

Using two NAND gates we may build a one-shot as in Figure 12.19. Here the trigger must be a

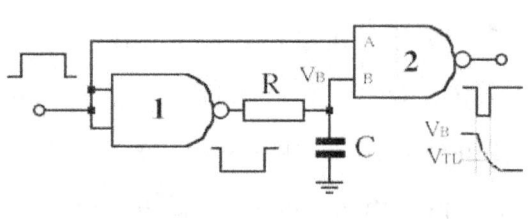

Figure 12.19

"high" pulse that lasts longer than the output "low" pulse. The stable state is "0" at input and "1" at output.

The rising edge of the trigger pulse toggles the first NAND (inverter), as well as the second NAND. The voltage $V_B(t)$, fed to the input B of the second NAND through the RC low-pass filter, decays exponentially with the law $V_B(t) = V e^{-t/RC}$. The toggling occurs again for $V_B(T) = V_{TL}$, where V_{TL} is the "low" threshold voltage. The output pulse width is therefore $T = RC \ln(V/V_{TL})$.

The examples of monostable circuit above described give an idea of the working principle of one-shot devices; however there are commercially available IC that implement monostable (e.g. 74121, 74122, 74123, 9602, 8853, 4538...) with added useful features, as free choice between rising or falling edge triggering, Q and $\overline{Q}$ output. These devices requires only external RC for setting the output pulse width.

12.5. Astables

Chapter 9 described several examples of astable multivibrators made by comparators plus RC negative feedback. Much more compact astable multivibrators can be made using logic ports, as shown in Figure 12.20, exploiting the delay provided by an RC low-pass filter that feeds the output back into the input.

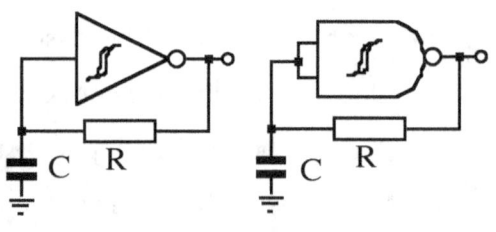

Figure 12.20

The symbol added inside the gates in Figure 12.20 marks

the gate hysteresis[54] due to *Schmitt trigger* inputs. The Schmitt trigger is simply a comparator with hysteresis [55].

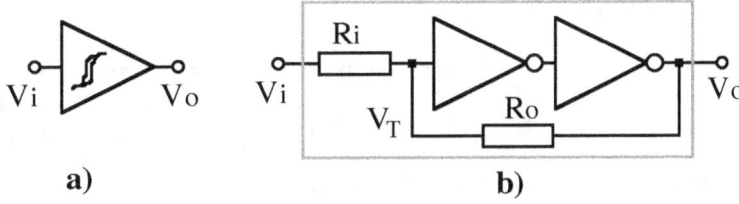

Figure 12.21

A normal buffer (a non-inverting gate that may be made of two inverters in series), may be transformed into a Schmitt trigger by using a resistive divider in a feedback loop, as shown in Figure 12.21b, where $R_i < R_0/2$. Let us consider a CMOS buffer biased between $V_{SS}=0$ V and $V_{DD}=5$ V, with threshold equal to $V_{DD}/2$. If the output is "low" ($V_0 \approx 0$ V) the toggling occurs when the input voltage V_i reaches the "high" threshold voltage V_{TH}. Because $V_T = V_i R_0/(R_i + R_0)$, we get $V_{TH} = (1 + R_i/R_0) V_{DD}/2$. The "low" threshold voltage V_{TL}, for $V_0 = V_{DD}$, is obtained by calculating V_T from superposition of V_i and V_0 sources and letting $V_T = V_{DD}/2$. From the relation $V_T = V_i R_0/(R_i + R_0) + V_0 R_i/(R_i + R_0)$, we get $V_{TL} = (1 - R_i/R_0) V_{DD}/2$. The hysteresis width is therefore $\Delta V = V_{TH} - V_{TL} = V_{DD} R_i/R_0$, that may be adjusted changing R_i or R_0.

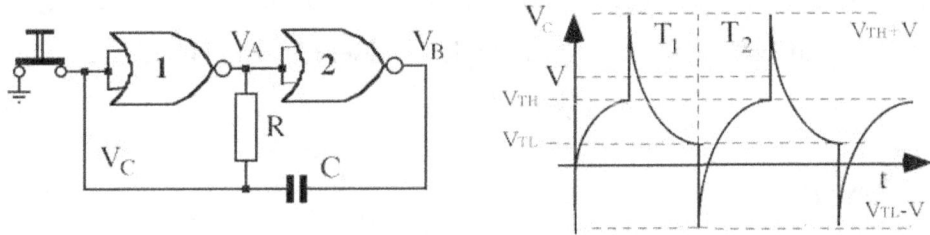

Figure 12.22

An astable multivibrator may be obtained from two inverters and an high-pass filter, as in Figure 12.22. The working principle is the following. The switch is initially closed, so that $V_C = V_B = 0$ and $V_A = V$ the capacitor is discharged. When the switch is opened, the capacitor starts charging through resistor R and the voltage V_C rises until it reaches the "high" threshold V_{TH}: at this time the output 1 toggles to "0" and output 2 toggles to "V", so that $V_C = V_{TH} + V$ and $V_A = 0$. The current across R changes sign and the capacitor decays with the time constant RC: $V_C(t) = (V_{TH} + V) \exp(-t/RC)$. After the time T_1 we the capacitor voltage reaches the "low" threshold: $V_C(t) = V_{TL}$. Solving for T_1 we get $T_1 = RC \ln[(V_{TH} + V)/V_{TL}]$. At this time $V_A = V$ and $V_C = V_{TL} - V$, the current again changes direction and the capacitor voltage follows the equation $V_C(t) = V + (V_{TL} - 2V) \exp(-t/RC)$, reaching the "high" threshold V_{TH} after the time T_2: where

---

[54] For example the hex schmitt trigger inverters (in the TTL family: 7414 and in the CMOS family: 4584) or the quad schmitt trigger NAND (in the TTL family: 74VH132 and in the CMOS family:4093).
[55] See chapt. § 9.2.

$T_2 = RC \ln[(2V - V_{TL})/(V - V_{TH})]$.

This circuit offers two output in phase opposition, but the signals are square wave only when the threshold voltages are symmetric with respect to the bias voltages (e.g. $V_{TL} = \Delta V$, $V_{TH} = V - \Delta V$), which holds for CMOS but not for TTL. A trick for adjusting separately the two time constants we may use the circuit shown in Figure 12.23, where the time constant is R_1C when $Q = "1"$ and R_2C when $Q = "0"$.

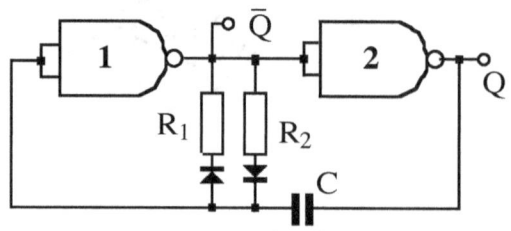

Figure 12.23

A simple multivibrator for high frequency square wave (up to several MHz, because the intrinsic delay τ due to the finite speed of signal transmission through the gates is of the order of some nanoseconds), is shown in Figure 12.24, for $2n+1$ inverters, with $n \geq 1$.

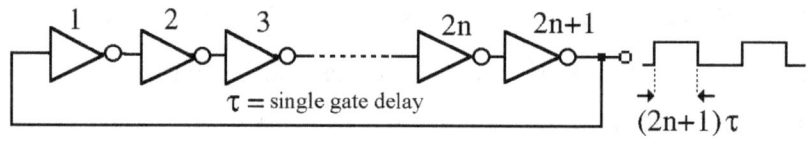

Figure 12.24

The first $2n$ inverters behave like the RC-filter delay in circuit of Figure 12.20. The square wave frequency in this circuit is not exactly predictable: it does depend on temperature and on bias voltage.

12.6. Monostable with delay

The delay generated by an odd chain of inverters may be used to build a monostable (one-shot) circuit as in the two examples shown in Figure 12.25.

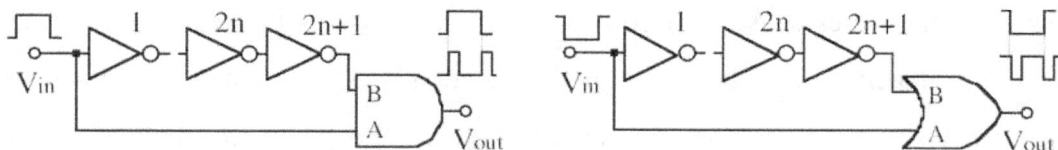

Figure 12.25

The pulse width is $T = (2n+1)\tau$. The stable state of the inputs in both NAND and NOR gates is complementary, because of the inverters chain; therefore the stable output is "low" in the NAND circuit and "high" in the NOR circuit. When the input voltage V_{in} changes state (either going "high" or going "low") immediately $A=B$, so that the output voltage V_{out} changes state. Only after the delay T also the input B changes state, thus toggling the output. If T_0 is the time interval between two transitions in the input voltage (the input pulse width) it must be $(T_0 > T)$.

12.7. Delay generator

A simple delay generator may be obtained with the circuits shown in Figure 12.26.

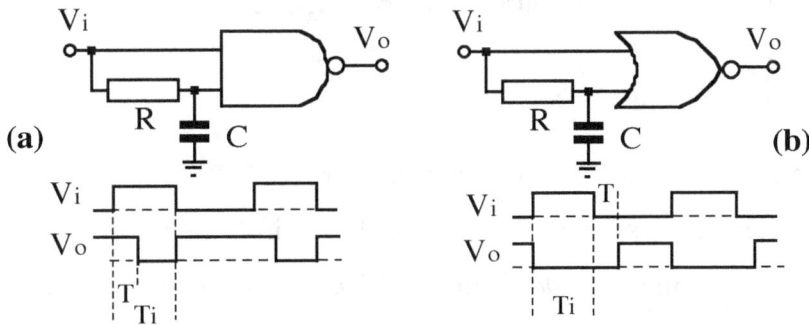

Figure 12.26

In the NAND version the output pulse falling edge is delayed with respect to the rising edge of the input pulse of the time interval $T = RC \ln[V/(V-V_{TH})]$. In the NOR version the output pulse rising edge is delayed with respect to the falling edge of the input pulse of the time interval $T = RC \ln(V/V_{TL})$. In both cases we must warrant $T < T_i$, where T_i is the input pulse width. By connecting these two circuits in series (with identical RC) the input pulse will be reproduced at the output with the delay T.

13. Some special IC

In this chapter we describe some popular IC that do not belong to the categories illustrated in the previous chapters: timers, IC voltage sources, analog switches.

13.1. The timer: a simplified description

The *IC timer* is made essentially by two comparators, one RS flip-flop, two transistors (one switch and one inverter). An essential drawing is shown in Figure 13.1, where also an external RC filter is connected to the *threshold* and *discharge* ports. The shown circuit behaves as monostable pulser

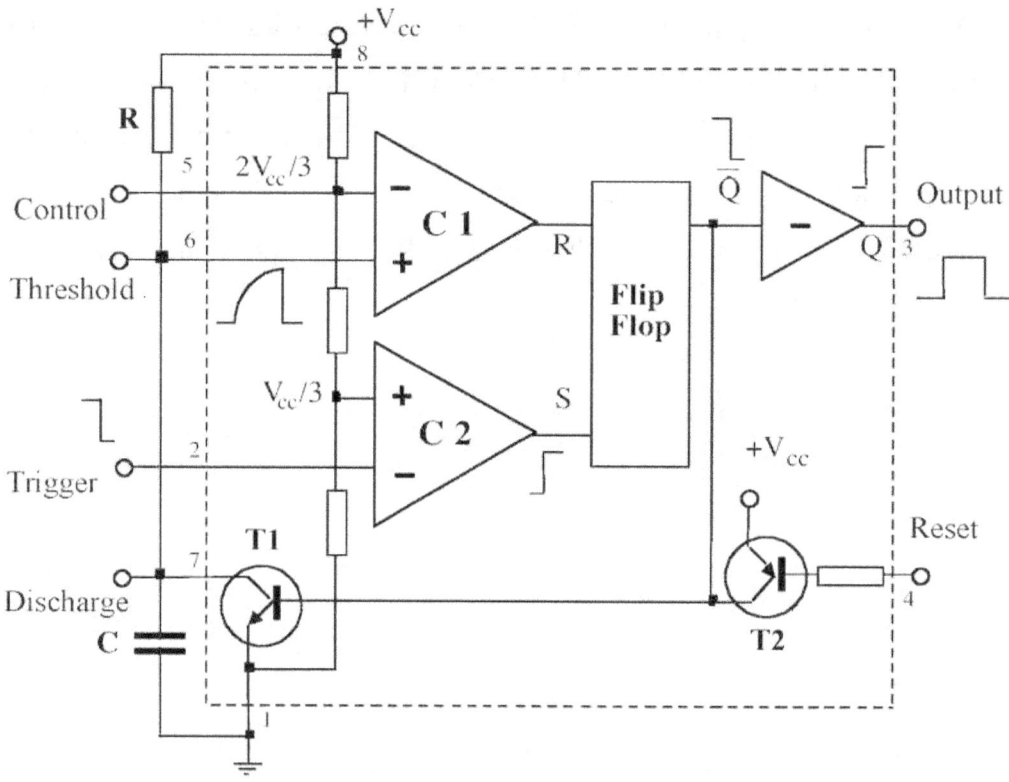

Figure 13.1

In the stable state the *trigger* port is kept at a voltage higher than $1/3 V_{cc}$, the transistor T1 is ON, the capacitor C is discharged, and both comparators have "low" output. When the trigger input falls below $1/3 V_{cc}$, the comparator C2 toggles, and the output pulse has width determined by the time constant RC. The pulse width is set by the time required to charge the capacitor C up to the voltage $2/3 V_{cc}$ through the resistor R. at this time the comparator C1 toggles, and the circuit reverts to the initial state.

The Flip-Flop $\overline{Q}$ output, normally "high", is forced "low" by the comparator C2 (signal S = *set*), and is forced "high" by the comparator C1 (signal R = *reset*). The transistor T1 is driven by the output $\overline{Q}$: it turns OFF when $\overline{Q}$ goes "low" (SET), and returns ON when $\overline{Q}$ goes "high" (RESET), thus discharging the capacitor.

The output may be forced "low" anytime by setting "low" the *reset* port that switches ON the transistor T2, which in turn switches ON transistor T1.

13.1.1. The timer 555

The most commonly used IC timer is 555 [56]. Its connection diagram is the one shown in Figure 13.1, where the value of the three resistors in the voltage divider is 5 kΩ, and the pin-out is resumed in Figure 13.2. The bias voltages are normally $+V_{cc} = (5 \div 15)\,V$ and $-V_{cc} = 0\,V$, but different values may be used, with a maximum voltage between pins 8 and 1 of 16V. For example we may use $+V_{cc} = +7\,V$ and $-V_{cc} = -7\,V$.

The output voltage (pin 3) in the "high" state is $+V_{cc} - 1.7\,V$ and $-V_{cc} + 0.3\,V$ in the "low" state[57].

The pin 5 (*Control Voltage*) is connected to the inverting input of comparator C1, and its voltage V_{CV} may be forced to a value different from the normal one ($V_{CV} = {}^1\!/_3\,V_{cc}$) When not used, this pin is frequently connected to $-V_{cc}$ through a capacitor to improve the immunity to noise.

The pin 2 (*trigger*) toggles C2 when its voltage crosses the value ${}^1\!/_3\,V_{cc}$, or the value ${}^1\!/_2\,V_{CV}$.

The minimum trigger pulse-width is 1 μs.

The pin 6 (*threshold*) is connected to the non-inverting input of comparator C1 and it forces toggling of C1 when its voltage crosses the value ${}^2\!/_3\,V_{cc}$, or the value ${}^1\!/_2\,V_{CV}$.

The pin 4 (*reset*) forces the output "low" when its voltage falls below the value $-V_{cc} + 0.7\,V$.

The pin 7 (*discharge*) is the open collector of a npn transistor (T1 switch).

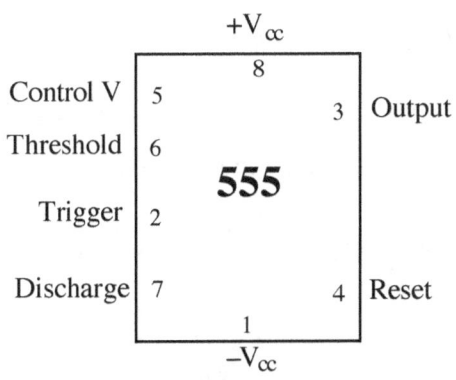

Figure 13.2

13.1.2. A monostable pulser made with 555 timer

We add to the circuit of Figure 13.1 an input capacitor C_i and a voltage divider (R_1, R_2) as in Figure 13.3, and we calculate the width of the output pulse produced by an input negative pulse V_T.

---

[56] This device, introduced in 1971 by Signetics as NE555, is now made by many companies in the original bipolar and also in low-power CMOS types (Exar XR555, Motorola MC1455, National LM555, Raytheon RM555, RCA CA555, Texas SN7255). The dual-type (two timers inside the same chip) is named 556. See also http://www.kpsec.freeuk.com/555timer.htm

[57] In the CMOS versions the output may reach $+V_{cc}$ and $-V_{cc}$.

Let us assume that input falling edge occurs at t=0, so that the capacitor voltage $V_C(t)$ starts from zero. It rises towards V_{cc} with exponential law: $V_C(t) = V_{cc}(1-\exp[-t/RC])$. The comparator C1 toggles at the time t=T when $V_C(T) = {}^2/_3 V_{cc}$, that gives $T = RC\,\ln(3)$ for the output pulse width. The divider (R_1,R_2) biases V_T so that $V_T > {}^1/_3 V_{cc}$: R_2 may be removed if the amplitude of the falling edge is larger than V_{cc}.

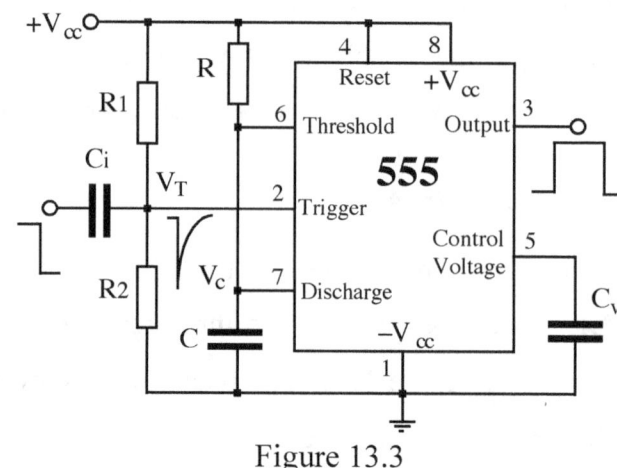

Figure 13.3

For example with $R_1 = R_2 = R_i = 10\,k\Omega$, we get $V_T(0) = {}^1/_2 V_{cc}$, and a time constant $R_i C_i / 2$ for the input high-pass filter. With a falling edge amplitude equal to ${}^1/_3 V_{cc}$, the time evolution of V_T is $V_T(t) = (V_{cc}/2)(1 - {}^2/_3 \exp[-2t/R_i C_i])$. Therefore the time interval τ^* in which $V_T < {}^1/_3 V_{cc}$ is $\tau^* = R_i C_i \ln(2)/2$, which must be $\tau^* > 1\,\mu s$: this set the minimum values of the capacitor $C_i > 100\,pF$.

On the other hand V_T must reach the stable value $V_T = {}^1/_2 V_{cc}$ before the end of the output pulse in order to avoid[58] *retriggering*, and therefore must be $C_i < 2(R/R_i)C$.

13.1.3. An astable pulser made with 555 timer

If we short the trigger pin to the threshold pin and we connect the discharge pin to the voltage divider (R_1,R_2) that charges the capacitor C, as in Figure 13.4, we obtain an astable pulser [59].

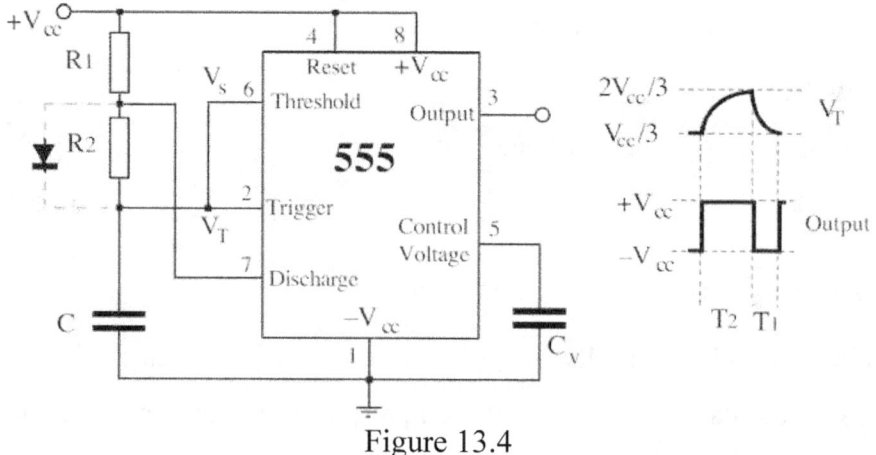

Figure 13.4

Let us start the analysis when the discharge pin is shorted to ground by the transistor T1: the capacitor C discharges through R_2, with time constant $R_2 C$, until $V_T = {}^1/_3 V_{cc}$. At this time the

---

[58] Alternatively the input pulse must be shorter that the output pulse.
[59] For a nice simulation of this circuit see http://www.williamson-labs.com/pu-aa-555-timer_slow.htm

comparator C2 toggles and the Flip-Flop switches-off T1. Now C is charged through R_1 in series with R_2, with time constant $(R_1+R_2)C$, until $V_S = 2/3 V_{cc}$. At this time the comparator C1 toggles switching-on T1, and the system reverts to the initial state.

In the first phase the time evolution of the voltage V_T (=V_S) is $V_T(t) = 2/3 V_{cc} \exp(-t/R_2C)$, that gives for the "low" output duration: $T_1 = R_2C \ln 2 \approx 0.693 R_2C$.

In the second phase ("high" output duration T_2) the time evolution of the voltage V_S (=V_T) is $V_S(t) = 1/3 V_{cc} + 2/3 V_{cc}\{1 - \exp[-t/(R_1+R_2)C]\}$, that gives $T_2 = (R_1+R_2)C \ln 2 \approx 0.693(R_1+R_2)C$.

The output cannot be a square wave; in fact the ratio $T_2/T_1 = 1 + R_1/R_2 > 1$ because the lower limit to R_1 is set by the maximum current tallowed for T1: $I_m = V_{cc}/R_1 \approx 100$ mA. However assuming $R_2/R_1 = 100$, the output asymmetry becomes only 1%. Another way to obtain a better Symmetry is by adding a diode in parallel to R_2, as shown in Figure 13.4. In this way we get in the second phase $V_S(t) = 1/3 V_{cc} + (2/3 V_{cc} - 0.6V)\{1 - \exp(-t/R_1C)\}$, accounting for the 0.6V bias voltage of the diode during capacitor charging. For example with $V_{cc} = 15$ V we get $T_2 \approx 0.76 R_1C$.

13.1.4. A square wave generator

A pure square wave generator may be obtained from a 555 timer as shown in Figure 13.5, where the capacitor is charged and discharged through the output port.

The resistor R_1 (non necessary in CMOS timers) is required in TTL timers to allow the output voltage reaching the value $+V_{cc}$, instead of $+V_{cc} - 1.7$ V. The half-period of the square wave is $T/2 = RC \ln 2$.

An auxiliary output signal (load R_L) is available at the discharge pin.

Note that the load R_L may be linked to any voltage: to $+V_{cc}$ as in Figure 13.5, or to any other value in the range ($+V_{cc}$, $-V_{cc}$), thus offering square wave with the desired amplitude. Moreover the load applied to output 2 does not affect the charge/discharge current of the capacitor.

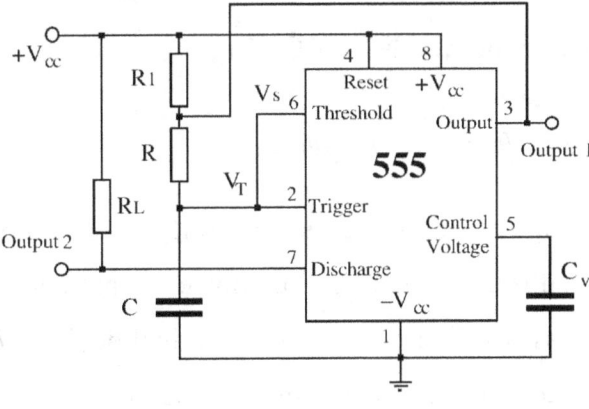

Figure 13.5

13.1.5. A linear voltage-to-frequency converter

In chapter 10, figure 10.15 shows a simple (*quasi-linear*) voltage-to-frequency converter made with two OA. Using one OA and one 555 timer CMOS[60] we obtain a *perfectly linear* voltage-to-frequency converter. In Figure 13.6 the OA is a differential integrator: $V_T(t) = (V_o - V_i)t/RC$, and the timer produces an output pulse V_o with width $\tau_o \approx 1.1 R_o C_o$.

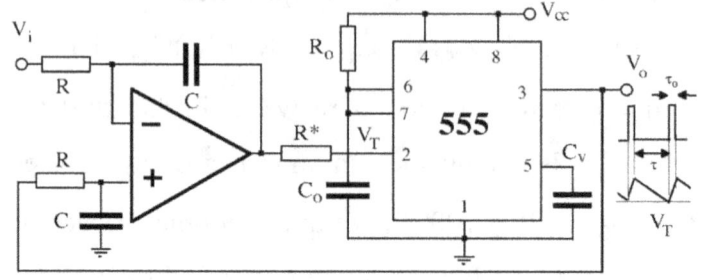

With a positive control voltage V_i the pulse frequency f is proportional to V_i:

$f = V_i / V_{cc} \tau_o$.

Let us assume that at some time $t = t^*$ the integrator output voltage is $V_T(t^*) > V_{cc}/3$, and the timer output is

Figure 13.6

$V_o = 0$: therefore V_T must decrease linearly with time: $V_T(t) = V_T(t^*) - V_i(t-t^*)/RC$, reaching the threshold voltage $V_{cc}/3$ at a time that we assume to be $t=0$. The output pulse begins ($V_o = V_{cc}$) and the differential integrator output linearly increases with the law: $V_T(t) = V_{cc}/3 + (V_{cc} - V_i)t/RC$. The pulse stops at the time τ_o, when $V_T(\tau_o) = V_{cc}/3 + (V_{cc} - V_i)\tau_o/RC$, so that V_T decreases reaching the threshold $V_{cc}/3$, at a time T, and the cycle is closed. During the negative ramp we have $V_T(t) = V_T(\tau_o) - V_i(t - \tau_o)/RC$, and setting $V_T(T) = V_{cc}/3$ we get $T = V_{cc}\tau_o/V_i$, so that the frequency is $f = 1/T = V_i / 1.1 R_o C_o V_{cc}$.

The time constant RC of the integrator does not affect the frequency, but its value is not arbitrary, because it does affect the slopes of the $V_T(t)$ ramps: the peak value V_p of $V_T(t)$ is $V_p = V_{cc}/3 + (V_{cc} - V_i)\tau_o/RC$, and it must be $V_p < 2V_{cc}/3$, so that, in the limit case $V_i \approx 0$ we must satisfy the condition $RC > 3\tau_o$.

13.2. IC voltage reference

Chapter 6 describes some voltage reference sources made with zener and OA with negative feedback. These circuits, however, are also commercially available as compact IC that may be classified in 5 classes: two-terminal devices (*band gap voltage reference*), *programmable zener*, three-terminal fixed-voltage sources, three-terminal adjustable regulators, and four-terminal adjustable regulators.

The band gap voltage reference are essentially zener with small temperature coefficient, down to

---

[60] Here the output toggles between $+V_{cc}$ and zero.

0.1 ppm/°C, with a reference voltage V_Z weakly dependent on current. The current available to the load is about $I_o = 10\,mA$, while the bias current is $I_p \approx 1\,mA$. Many values are available for V_Z, e.g.: 1.8, 2.0, 2.2, 2.4, 2.7, 3.0, 3.3, 3.6, 3.9, 4.3, 4.7, 5.1, and 5.6 V for LM103xx (where xx stays for the value V_z), 1.22 V for LM113, 1.2V for AD589, 6.95 V for LM199/299/399, and 6.9 V for LM129/329.

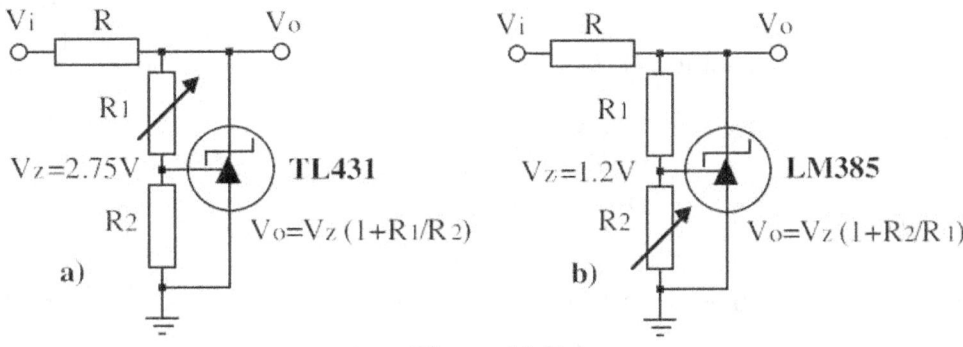

Figure 13.7

The programmable zener are three-ports devices that must be used as shown in the diagrams of Figure 13.7a, or Figure 13.7b, depending on the device type. Without voltage divider (R_1, R_2) the devices behaves as a normal zener.

The three-terminal fixed-voltage sources (Figure 13.8) generate a stable output voltage V_o (either positive or negative) in a wide range of input voltage V_i: from $V_i \approx V_o$ to $V_i \approx 10\,V_o$. With a minimum bias voltage of a few mA they may supply to the load currents up to 3 A, with a small temperature coefficient (10÷30 ppm /°C) for the

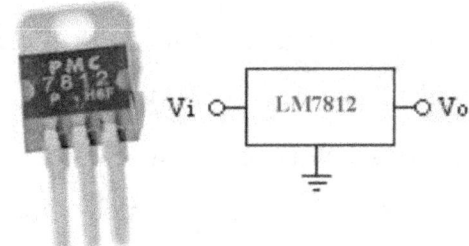

Figure 13.8

output voltage V_o. Typical values of output voltage are: +2.5 V (AD580, AD1403), +5 V (LM123/223/323, LM109/209/309, AD581), and −5 V (LM145/245/345).

Low power models offer more values: (typically V_o = 5, 6, 8, 10, 12, 15, 18, 24 V): LM140/240xx, LM341xx, μA78Mxx, LM78xx (for positive V_o) and LM120/220/320xx, LM79xx, μA79Mxx (for negative V_o), where xx stays for the V_o value. E.g.: μA79M05 for −5 V, LM22018 for +18 V.

The *3-terminal adjustable regulator* typical wiring is shown in Figure 13.9. The output voltage V_o ranges from 1.25 V to V_i−2 V, where the input voltage V_i is normally limited to 35V (40V in some models). The value of resistor R_2 may go down to zero (for minimum $|V_o|$).

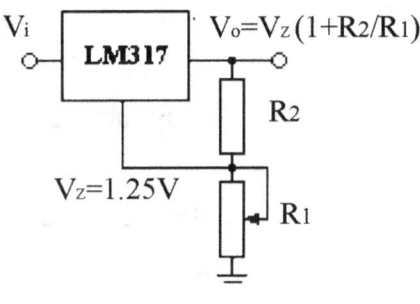

Figure 13.9

There are model for positive output (LM150/250/350, LM117/317,TL317) and for negative output (LM137/237/337).

The wiring in the *4-terminal adjustable regulator* is similar, but here the role of the two resistors in the voltage divider is exchanged (see Figure 13.10): here the minimum output is for $R_1=0$.

For example in the positive output μA78G we get $V_z=+5$ V, and in the negative output μA79G we get $V_z=-2.2$ V.

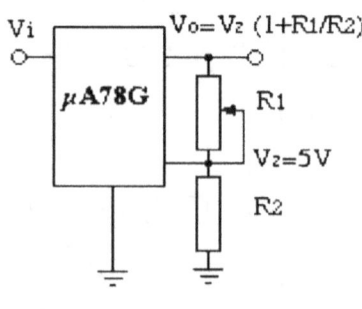

Figure 13.10

13.3. Analog switches

The ideal switch may be defined as a bi-stable two-terminal device that an external action may toggle between zero resistance R_{on} and infinite resistance R_{off}.

The external command may be mechanic (e.g. manually operated switch) or electro-mechanic (relays) or simply a voltage signal (analog-switch).

The real switch differs from the ideal one because the resistance R_{on} in the "closed" state is not zero and the resistance R_{off} in the "open" state is not infinite. In the analog switches may be $R_{on} > 100\,\Omega$ and $R_{off} < 100\,k\Omega$.

The advantages of analog switches are mainly their speed, and the possibility of use low-power command signals. Analog switches may be implemented with bipolar transistors or with FET (typically CMOS). In the first case the current must flow through the two terminals of the switch in a given direction (*unipolar* switch), in the second case in both directions (*bipolar* switch, i.e. the two terminals may be interchanged).

There are many commercially available IC analog switches, with various configurations: double, quad or even more switches integrated inside a single chip.

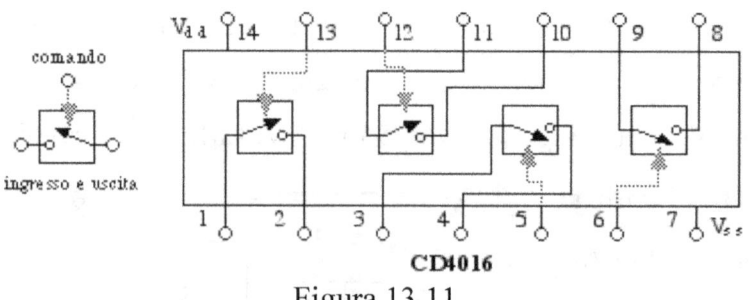

Figura 13.11

One of the most popular model is 4016 [61] (CMOS-Quad-Bilateral-Switch) whose block diagram is shown in Figure 13.11.

It must be biased by a maximum voltage ($\Delta V = V_{DD} - V_{SS}$) in the range from $+3$ V up to $+20$ V but all terminals (included command pins), cannot go lower than $V_{SS} - 0.5$ V or higher than $V_{DD} + 0.5$ V. The maximum current is 10 mA. Typical value of R_{on} is 300 Ω, and the

---

[61] CD4016 produced by RCA, or 74MM4016 from National, or 4066 with $R_{on} \approx 90\,\Omega$.

leakage current in the "open" state is of the order of fractions of nA.

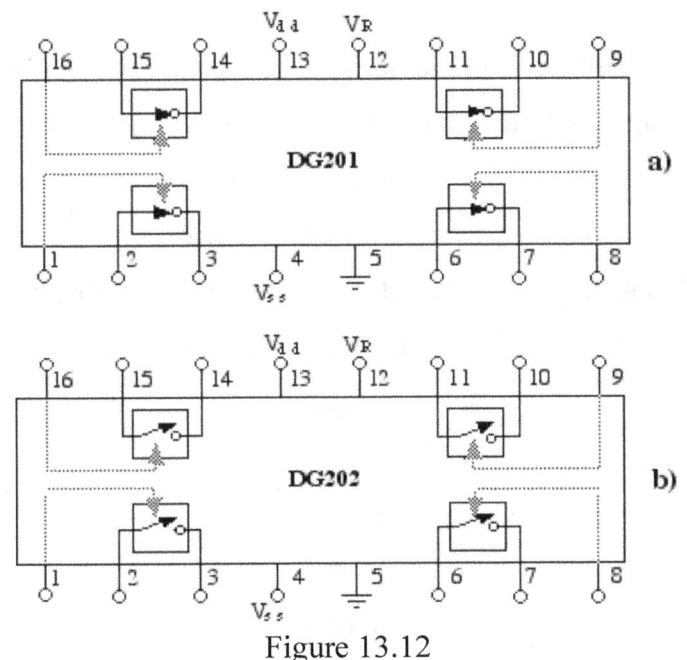

Figure 13.12

More sophisticated CMOS quad bilateral switches are the models 201 and 202[62]. The block diagram is shown in figures 13.12a and 13.12b, respectively. The first type has the 4 switches normally closed (with command voltage is "low"), while the second type has the 4 switches normally open..

These IC have dual power supply, symmetric and referred to ground, with values in the range from ± 5 V and ± 18 V. The command threshold voltage ranges from $+0.8$ V and $+2$ V : e.g. $V_{DD} = +15$ V the threshold is $+1.4$ V. The threshold voltage may be adjusted through the V_R pin.

The maximum current may be higher that 20 mA, with $R_{on} \approx 60\,\Omega$, and leakage currents of fractions of nA.

The different chips are frequently identified by acronyms that define the functions: SPST means Single Pole Single Throw, QPDT means Quad poles Double Throw, and so on... (see Figure 13.13

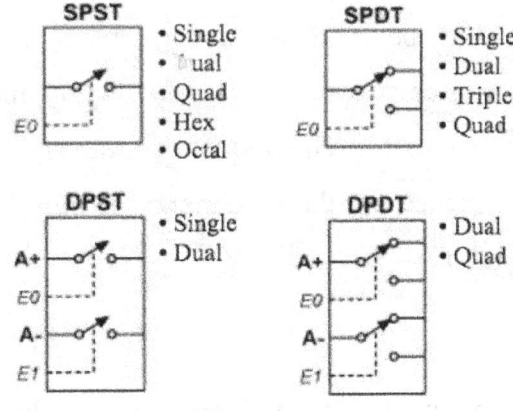

Figure 13.13

---

[62] DG201 from Siliconix, or Maxim, or equivalent SW201 from Precision Monolitics and LF11201 from National, (and DG202, or equivalent SW202 and LF11202).

14. Transducers, sensors and interfacing techniques

The name *transducer* defines a device that transforms a signal expressed in a physical quantity (temperature, velocity, magnetic field...) into a signal expressed in a different physical quantity. Transducers are usually divided into two classes: *sensors* and *actuators*: The name *sensor* defines a device that converts the value of a physical quantity, or its changes into an electrical signal. The name *actuator* defines a device that converts electrical signals into changes of some physical quantity. Some transducers are reversible: they may be used either as sensors or as actuators.

actuator		*sensor*
magnetodynamic loudspeaker	⇐⇒	microphone
d.c. electric motor	⇐⇒	dynamo tachymetric
LED diode	⇐⇒	photodiode
piezo-transducer	⇐⇒	piezo-sensor

Table 14.1

The term *interfacing* is used for the techniques used to transform the signal generated by a sensor into an electrical signal, or to adapt the amplitude and shape of the signal to required features, or to generate a suitable signal to drive a given actuator.

In this chapter we will analyze only some of the many existing actuator/sensors, to give a general idea of the simplest interfacing techniques. We well consider, as examples, transducers for four physical quantities: temperature, force, light, and position.

The *temperature transducers* may be used as thermometers, but also as level sensors, flux sensors, thermal conductivity sensors, ... The *force transducers* may also be used as pressure sensors, as sound generators/sensors as, ... The *optical transducers*, depending on the wavelength may detect/generate visible light, measure the flux/energy of light beams, or X-rays, or may be used as thermometers (bolometers) ...

General features of a sensor

- *sensitivity* (ratio between the output signal and the change of the measured physical quantity)
- *resolution* (minimum change of the input quantity that can be detected)
- *accuracy or precision* (maximum error affecting the measurement)
- *range* (range where the measurement may be performed with the given accuracy)
- *non-linearity* (departure of the transfer function from linear behavior)
- *hysteresis* (non-reproducibility of the transfer function after large changes)
- *dynamic characteristics* (response time, rise-time, settling-time, damping, band-pass width)
- *signal/noise ratio* (due to internal noise or pick-up noise)
- *output impedance* (in series for voltage source, in parallel for current source)
- *drift* (thermal, aging)

14.1. Temperature sensors

Temperature sensors may be divided in three broad classes: resistive sensors, diodes, and thermocouples.

14.1.1. Resistance thermometers

The resistive temperature detectors (RTD may be metals, semiconductors or carbon-resistors. The metallic RTD are usually made of copper, nickel or platinum. The platinum RDT are the most reliable because a Pt wire may be produced with very small impurity content, which makes the temperature coefficient of the sample highly reproducible (but they are very expensive).

The resistivity of a pure metal follows approximately (at temperatures not too low) the linear law $\rho(T) = \rho_0 (1+\alpha T)$, where ρ_0 is the residual resistivity at $T \approx 0\,K$, proportional to the impurity and lattice imperfections density, and $\alpha = (\partial R/\partial T)/R$ is the temperature coefficient: for platinum $\alpha \approx 3.85 \cdot 10^{-3}\,K^{-1}$, for copper $\alpha \approx 3.9 \cdot 10^{-3}\,K^{-1}$, for nickel $\alpha \approx 5\text{-}7 \cdot 10^{-3}\,K^{-1}$.

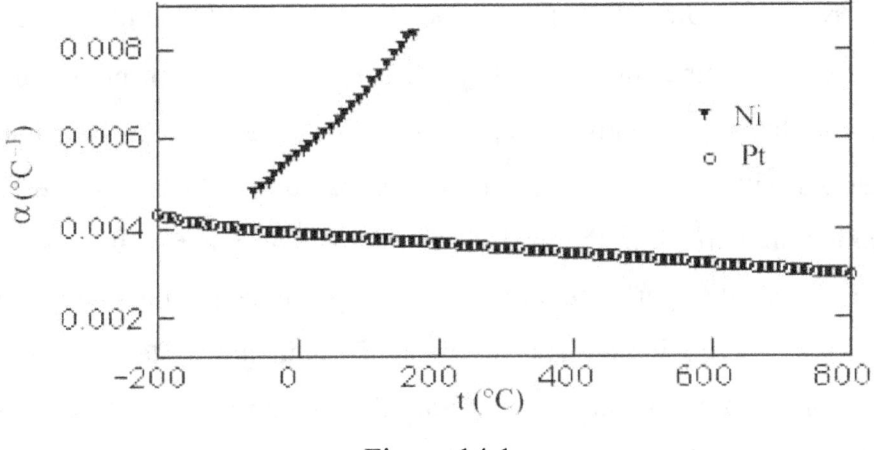

Figure 14.1

Metallic RDT have small mass (and therefore fast response) and good linearity over a large temperature range. They must be biased by a constant d.c. or a.c. current. Sensors with small dimensions have low electric resistance (typically $100\,\Omega$ at room temperature) and this impose some care in the interfacing technique in order to make negligible the error due to the cables resistance. Their sensitivity is limited by the Joule self-heating, which requires reducing the bias current and therefore the signal amplitude. Typical useful ranges: platinum from $10\,K$ to $800\,K$, nickel from $-60\,°C$ to $+300\,°C$ and copper from $-70\,°C$ to $+150\,°C$.

The simplest method to measure a resistance R_x is the voltamperometric method: we measure the voltage drop V_x across R_x due to the known current I_p flowing through it. By keeping constant I_p the R_x measurement reduces to V_x measurement. This technique may be accurate in the 4-terminals configuration where the bias cables are different from the voltage-detection cables

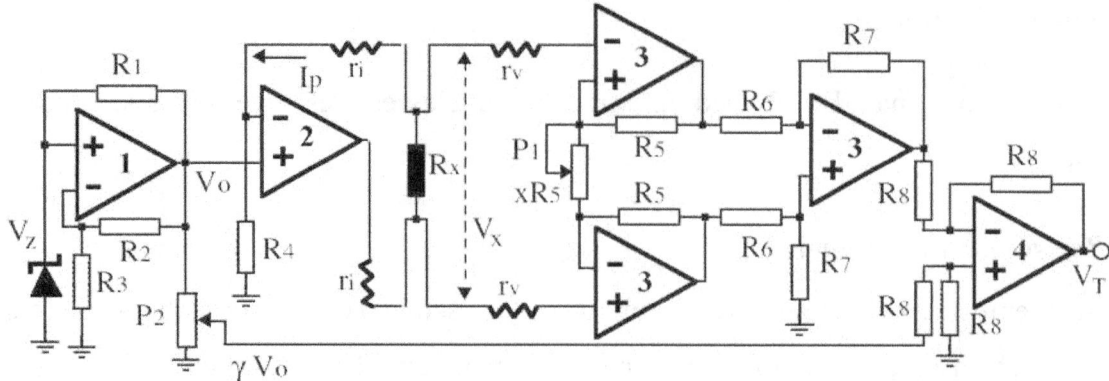

Figure 14.2

A simple interfacing circuit that uses the 4-terminals configuration is shown in Figure 14.2, where OA1 supplies a stable voltage reference[63] $V_o = V_z(1+R_2/R_3)$, OA2 provides the constant current[64] $I_p = V_o/R_4$. The signal $V_x = R_x I_p$ across the thermometer is measured by the *instrumentation amplifier* made of OA3,[65] with adjustable gain $G(x) = (1+2/x)R_7/R_6$. The scale factor dV_x/dT is set by the potentiometer P_1 that controls $G(x)$, while the scale origin is set (through the differential amplifier OA4) by the potentiometer P_2 that controls the fraction γ of the reference signal from the output signals: $V(T) = G(x)I_p R_x(T) - \gamma V_o$. This circuit allows reading the temperature of the body thermally anchored to R_x in kelvin, Celsius, Fahrenheit, or its temperature changes with respect to a reference temperature. The resistances r_i and r_v of the bias/detection cables are explicit in Figure 14.2: the voltage drop across the resistances r_v is made negligible by the very small input current of the high impedance instrumentation amplifier.

The circuit of Figure 14.2 with d.c. bias cannot distinguish real temperature signal, due to $R_x(T)$ changes from offset voltages of the amplifier chain. This problem may be avoided replacing the d.c. reference V_o with an a.c. *stable* signal.

An alternative method, that does not require a stabilized a.c. source, compares R_x with calibrated resistors in a Wheatstone bridge-configuration as in the circuit of Figure 14.3 where the bridge is biased by the sinusoidal voltage V[66].

---

[63] See chapter 6.
[64] See chapter 7.
[65] See chapter 4, §4.
[66] See chapter 10.

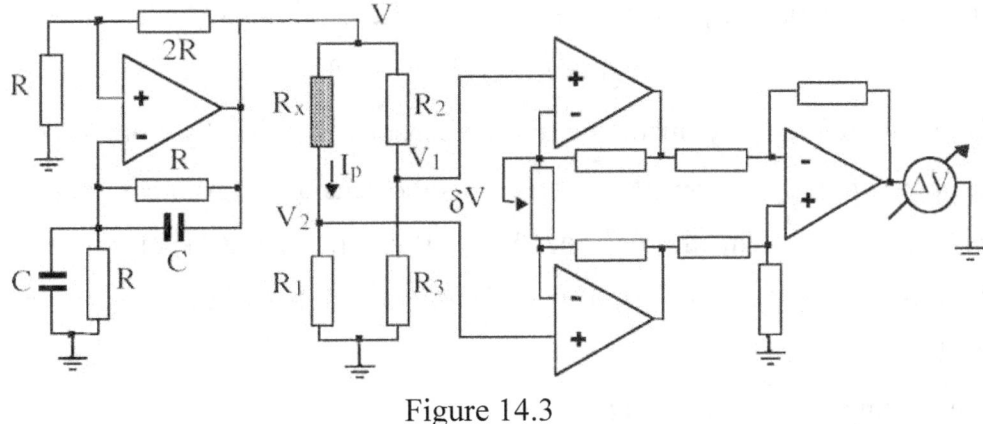

Figure 14.3

Here the bridge is balanced for $R_x/R_1 = R_2/R_3$. For example, assuming $R_2 = R_3 = R$ and using for R_1 a set of calibrated resistors (*decade resistor box*), the measurement is performed by adjusting the value of R_1 until the output *error signal* $\Delta V = G\delta V$, is minimized. This gives $R_x \approx R_1$. The value of the current I_p does not enter the balance equation, therefore we do not need a stable a.c. bias voltage. : If $R_1 = R_x(1+\varepsilon)$, with $\varepsilon = (R_1 - R_x)/R_x$ the error signal $\delta V = V_2 - V_1$ may be written $\delta V = RI_p(\varepsilon/2)/(2+\varepsilon)$. This equation shows that the error signal is linear only for very small values of the unbalance parameter ε.

This interfacing technique is frequently used with non-linear RTD and with high resistance RTD (that make negligible the cable resistances), such as semiconductor RDT. Semiconductor RTD (normally named *thermistors*) may have negative (NTC) or positive (PTC) temperature coefficient, depending on the dopant level and on the temperature range (Figure 14.4).

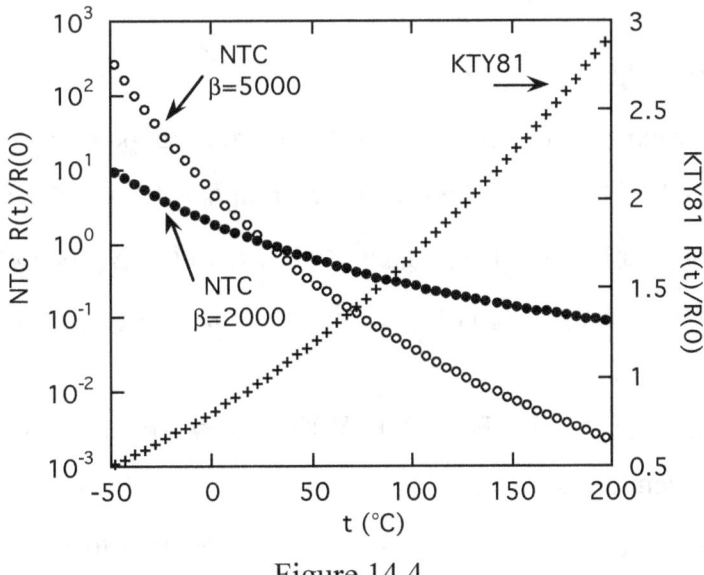

Figure 14.4

NTC *thermistors* have normally an exponential temperature dependence $R(T) = R_o \exp(-B/T)$, which implies high sensitivity and non-linearity ($\alpha = \partial R/R\partial T = -B/T^2$). The advantage of these sensors is the small size and the wide range of resistance value.

The thermistor characteristic equation is frequently written:

$$R(T) = R(T_0)\, e^{\beta(1/T - 1/T_0)},$$

where $R(T_0)$ is the reference temperature and the constant β (typically from 2000 to 5000 K) is named *characteristic temperature*, which measures the sensitivity. Another equation, commonly used, is the Steinhart-Hart equation[67]: $T = 1/\{A_1 + A_2[\ln(R_T)] + A_3[\ln(R_T)]^3\}$ where A_1, A_2, A_3, are parameters provided by the thermistor manufacturer, and R_T is the thermistor resistance at temperature T.

A technique to improve the linearity of the response in a bridge[68] (within a limited temperature range: $T_1 < T < T_2$) is shown in Figure 14.5. Note that this technique gives an output voltage *increasing* with temperature.

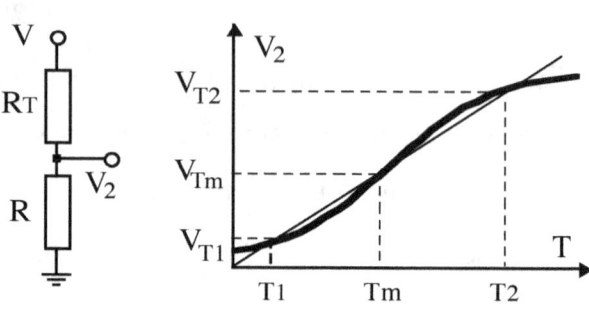

Figure 14.5

The divider shown in Figure 14.5 gives: $V_2 = V \cdot R/(R + R_T)$, where R_T is the NTC thermistor resistance at the temperature T; for $T \to 0$, $R_T \to \infty$ and $V_2 \to 0$, while for $T \to \infty$, $R_T \to 0$ and $V_2 \to V$.

Therefore we may choose for R a value such that $V_2(T_2) - V_2(T_m) = V_2(T_m) - V_2(T_1)$. Solving this equation we find the best value: $R = (R_{T1} R_{Tm} + R_{T2} R_{Tm} - 2 R_{T1} R_{T2}) / (R_{T1} + R_{T1} - 2 R_{Tm})$. As a first approximation a good choice is $R = R_{Tm}$.

14.1.2. Diode thermometer

The diode thermometer exploits the *quasi-linear* temperature dependence of the *forward voltage* V of a p-n junction, for T > 30 K when the flowing current I_d is kept constant.

In fact the diode characteristic curve is $I_d \approx I_o\, e^{qV/K_BT}$, where $I_o \approx A e^{-E_g/K_BT}$, K_B is the Boltzmann constant, q is the electron charge, E_g is the semiconductor energy gap and A is a constant that depends on the junction area[69].

We get $\ln I_d \approx \ln I_o + qV/K_BT = \ln A - E_g/K_BT + qV/K_BT$, or $V - E_g/q = -(K_BT/q) \ln(A/I_d)$, which is the linear dependence mentioned above: $V = V_o - \gamma\{I_d\} T$.

The voltage $V_o = E_g/q$ is the diode forward voltage extrapolated to 0 kelvin and γ is the slope which depends logarithmically on I_d, and decreases with increasing I_d. The advantages offered by

---

[67] See http://en.wikipedia.org/wiki/Steinhart–Hart_equation
[68] See http://mathscinotes.wordpress.com/2011/07/22/thermistor-mathematics/
[69] See also Appendix A.1 and § 8.4.1

this sensor is the linearity and the constant high sensitivity (about 2 mV/K).

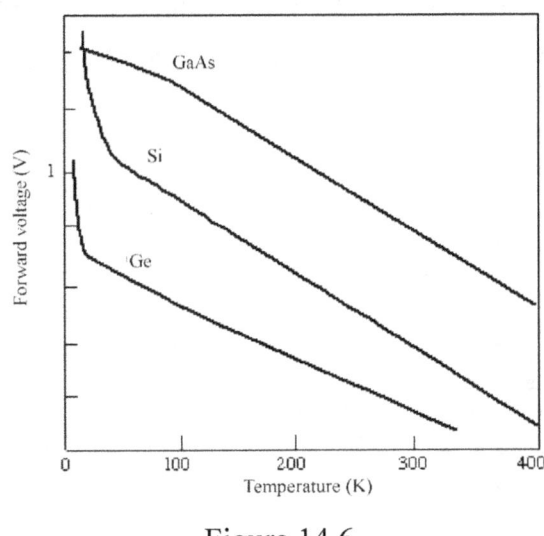

Figure 14.6

The diode thermometer requires a constant current bias ($I_d \approx 1 \div 100\,\mu A$): a.c. current cannot be used. A circuit suitable for diode thermometry is that shown in Figure 14.2 (obviously providing forward bias to the diode). A simpler circuit is shown in Figure 14.7, where the potentiometer P_1 adjusts the bias current I_d and the output voltage is $V_{out} = G[V_o - E_g + \gamma(I_d)T]$, where $V_o = VR/(R+P_2)$ and $G = R_o/R_1$ is the differential amplifier gain. The potentiometer P_2 provides the zero-scale adjustment.

There are commercially available sensors (as National LM335 or Texas STP35) that give an output voltage of 2.73 V at 0 °C, with temperature coefficient of +10 mV/K. These are IC that include with the sensing diode also the interfacing circuitry. Other models as Analog Devices AD590, and AD592, when biased by a voltage in the range from 4 V to 30 V, give an output current proportional to absolute temperature with temperature coefficient 1 µA/K.

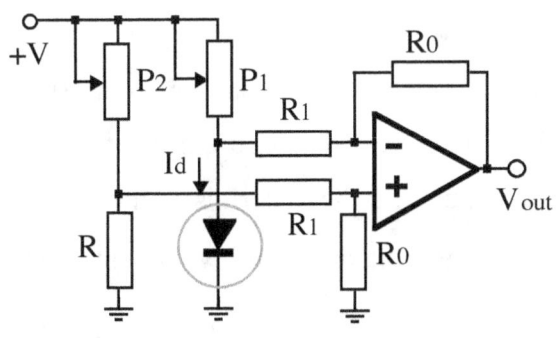

Figure 14.7

The working range is −55 °C +155 °C for AD590 and LM335, −25 °C +105 °C for STP35, and −25 °C +105 °C for AD592.

14.1.3. The thermocouple

The thermocouple is a temperature sensor that exploits the temperature dependence of the electromotive force (emf) in a junction of two different metals (Seebeck effect) [70]. This emf V_{TC} is an increasing function of T, almost linear near room temperature with a temperature coefficient $\partial V_{TC}/\partial T$ of the order of a few µV/K.

The main advantages of these sensors are: 1) speed, due to small mass; 2) easy thermal coupling; 3) extended working range, from 10 K to 1000 K; 4) low cost; 5) no bias needed [71]. Drawbacks:

---

[70] For the Seebeck effect see http://en.wikipedia.org/wiki/Thermoelectric_effect
[71] For accurate measurements, non only the sensor but also the wires connecting the sensor to the interfacing circuit must be thermally coupled to the sample, in order to avoid temperature gradients between the sample and the

non-linearity and low sensitivity. The sensitivity depends on the junction materials: the mostly used types are **J**: (Iron+, Constantan–) and **K**: (Cromel+, Alumel–), where Constantan is an alloy 60%Cu-40%Ni (also 55%Cu-45%Ni), Cromel is 90%Ni-10%Cr, and Alumel is 95% Ni-2%Mn-2%Al-1%Co.

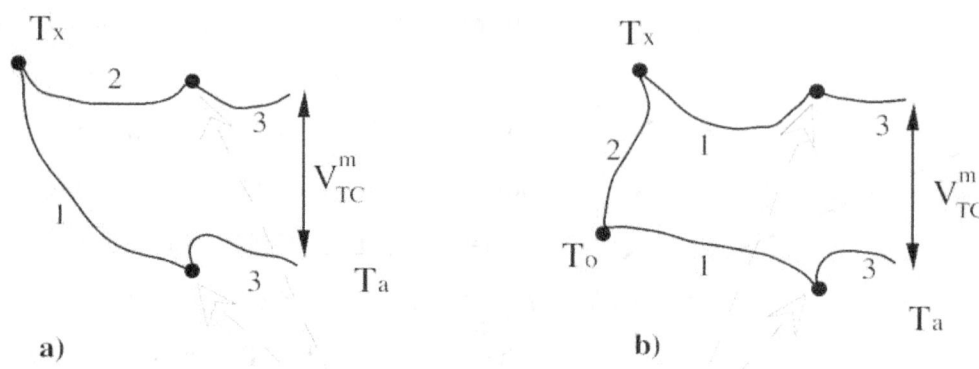

junctions with voltmeter wires

Figure 14.8

Note that when two wires of a thermocouple are connected to the interface terminals (usually made of copper), two more junctions (usually at room temperature T_a) are made (see Figure 14.8a): therefore any measurement of V_{TC} is the sum of three V_{TC}.

Let us name 1, 2 the materials of the two thermocouple wires and 3 the material of the voltmeter terminals and the *measured* $V_{TC}^{ab}(T_x)$ the emf of the junction between a and b, at temperature T_x; we get: $V_{TC}^m(T_x) = V_{TC}^{31}(T_a) + V_{TC}^{12}(T_x) + V_{TC}^{23}(T_a)$.

Because $V_{TC}^{23}(T_a) + V_{TC}^{31}(T_a) = V_{TC}^{21}(T_a) = -V_{TC}^{12}(T_a)$, we get $V_{TC}^m(T_x) = V_{TC}^{12}(T_x) - V_{TC}^{12}(T_a)$. Once known the function $V_{TC}^{12}(T)$ for each value T, we only need to measure T_a and to measure $V_{TC}^{12}(T_x)$ to obtain T_x.

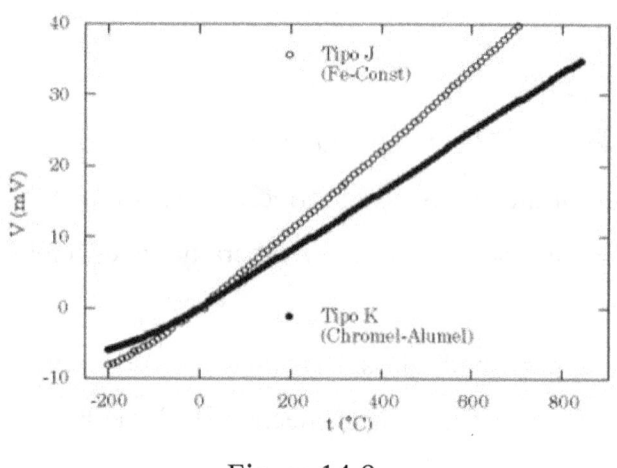

Figure 14.9

To avoid the measurement of the room temperature, we may add a *reference junction* (kept at a fixed known temperature T_o, e.g. 0 °C ice-bath), so that the wires loop is closed at T_a with the same metal (1-2-1, in Figure 14.8b) and the contributions $V_{TC}^{31}(T_a) = -V_{TC}^{13}(T_a)$ cancel out. Therefore the value of T_a does not affect the measured $V_{TC}^m(T_x) = V_{TC}^{12}(T_x) - V_{TC}^{12}(T_0)$.

---

sensor. Thermal coupling may be achieved by pressing the sensor against the sample (and using suitable oil/grease or glue with high thermal conductivity).

This quantity is published in the *thermocouple data tables* for a given reference temperature T_0.
In place of the usual ice-bath for the reference junction, the room temperature changes may be accounted for by an electronic automatic correction provided by a diode thermometer. An example of this approach is shown in Figure 14.10. The differential amplifier A1 must have high input impedance in order to make negligible the thermocouple wires resistance changes, and a high gain G because the source signal is of the order of few mV: with type **J** changes of 0.1 mV correspond to temperature changes of about 2 degrees.

With $R_o = R_a = 100\,k\Omega$, $x = 1/9$ and $R_i = 10\,k\Omega$ we get $G = 200$, i.e. a sensitivity of about 10 mV/K for this thermometer. The reference junction temperature compensation is achieved by injecting the signal V_5 into the inverting input of A1: which is transferred to the output with unity gain [72]. The signal V_5 is the sum of signal

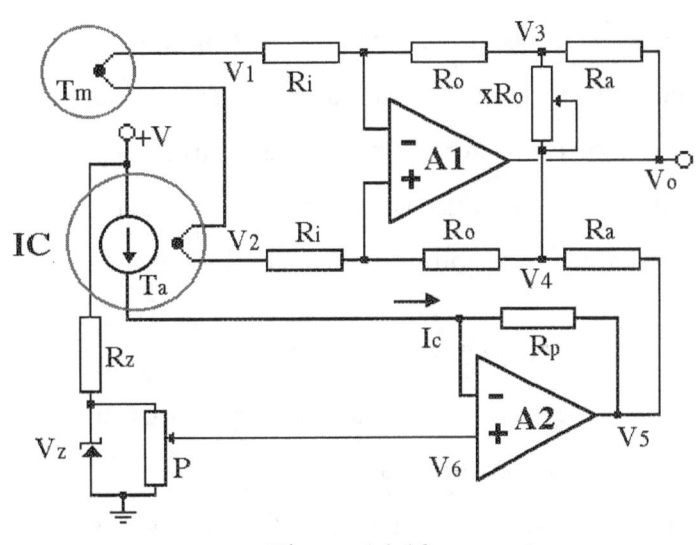

Figure 14.10

V_6 and signal $-I_c R_p$ produced by the calibrated current a $I_c(T_a)$ generated by the IC thermometer (e.g. AD590, AD592). The resistor R_p must be selected in order to compensate with $\partial V_5/\partial T$ the changes $\partial V_2/\partial T$, generated by the junction at room temperature. With $G = 200$, an IC thermometer sensitivity $\partial I_c/\partial T = 1\,\mu A/K$ and an output drift due to reference-junction $\partial V_5/\partial T = 10\,mV/K$, we must choose $R_p = 10\,k\Omega$. The potentiometer P allows adjusting the fraction V_6 of the stabilized voltage V_z (e.g. for kelvin scale we set $V_6 = 2.73\,V$ at $0\,^\circ C$)

There are commercially available IC (Analog Devices: AD 594 for J-type, AD595 for K-type) that include all the circuit of Figure 14.5, plus a TTL-alarm output that toggles when the thermocouple loop is opened.

14.2. Force and pressure sensors

The force sensors measure the deformation of an elastic object subject to the applied force: the elastic constant relating force and deformation is determined by calibration with known force values. The sensing object may be a piezoelectric crystal or a resistive bridge obtained from a

---

[72] See § 4.1, where V_5 replaces the 0 voltage at one end of resistor R''_a.

semiconducting wafer or the flexible electrode of a capacitor, or any elastic object connected to any suitable strain-detector (e.g. optical or magnetic). When the measured force is due to the collisions of a gas molecules against the sensing object we get a *pressure sensor*.

The force sensors measure the deformation of an elastic object subject to the applied force: the elastic constant relating force and deformation is determined by calibration with known force values. The sensing object may be a piezoelectric crystal or a resistive bridge obtained from a semiconducting wafer or the flexible electrode of a capacitor, or any elastic object connected to any suitable strain-detector (e.g. optical or magnetic). When the measured force is due to the collisions of a gas molecules against the sensing object we get a *pressure sensor*.

Many force sensors (usually named strain gauges) are made of a metallic or semiconductor resistors (wires or films) whose resistance is strain-dependent: the strain produces changes in the object geometry (e.g. a metal bar under tension becomes longer and thinner, so that its resistance increases, under compression its resistance decreases).

14.2.1. Piezoresistive pressure sensor

Most of today's pressure transducers consist of a four-piezoresistor[73] Wheatstone bridge fabricated on a single monolithic die using bulk-etch micromachining technology. The piezoresistive elements integrated into the sensor die are located along the periphery of the pressure-sensing diaphragm at the points appropriate for strain measurement: the diaphragm deformation, due to the applied pressure, changes the values of the 4 resistances and the output of the unbalanced bridge is a differential signal proportional to the bias voltage and to the applied pressure[74].

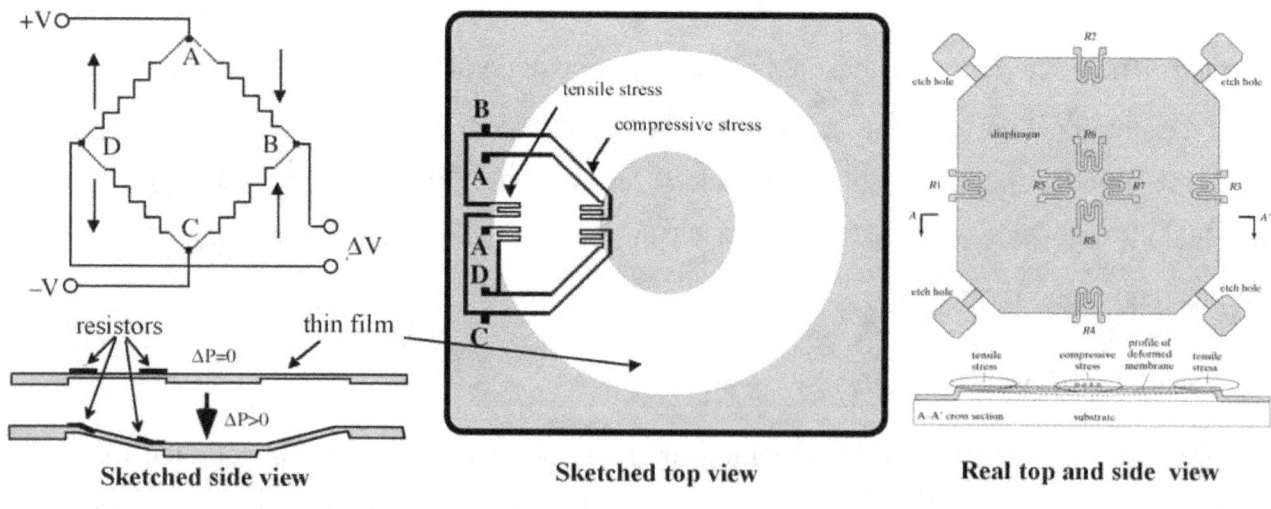

Figure 14.11

---

[73] See also http://en.wikipedia.org/wiki/Piezoresistive_effect
[74] For an extended description see http://www.mech.northwestern.edu/FOM/LiuCh06v3_072505.pdf

Figure 14.11 shows some details of a typical piezoresistive pressure sensor. The four resistors are shaped usually with a serpentine-pattern to increase resistance and sensitivity: resistors AB and CD work in compressive strain, while resistors BC and DA work in tensile strain, so that the bridge sensitivity is doubled.

In a sensor obtained from semiconductor wafer the resistors ($R \approx 5\,k\Omega$) must be biased by d.c. voltage (typically 10V). For absolute pressure sensors the full scale may reach 5 MPa, and for relative pressure sensors ranges from some Pa to some MPa [75].

In absolute sensors the diaphragm seals a small evacuated volume, while in relative sensors the reference pressure is the atmospheric pressure, or it may be different when measuring differential pressures. The sensitivity σ depends slightly on temperature: $\partial\sigma/\sigma\partial T \approx -10^{-3}\,K^{-1}$, as well as the offset ($\partial V_{os}/V\partial T \approx 10^{-4}\,K^{-1}$) so that some IC pressure sensors include temperature compensation circuitry. Sensitivity may be adjusted by trimming the bias voltage.

In Figure 14.12, the (Siemens KPY32) pressure sensor is biased by a voltage divider made by two resistors R_1 and a (Siemens KTY 10) PTC thermistor thus increasing the bias voltage with temperature.

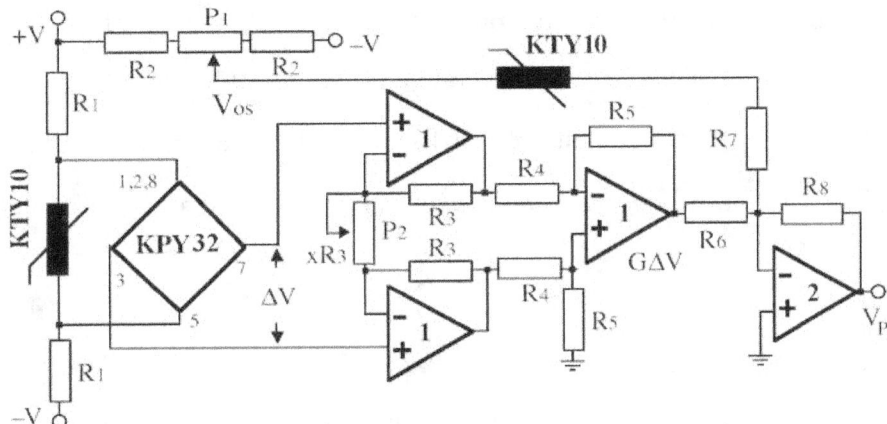

Figure 14.12

The gain of the instrumentation amplifier 1 is adjusted by potentiometer P_2, and the offset is zeroed by P_1, and temperature-compensated by the PTC in series to the resistor R_7 which scales the $-V_{os}$ signal at the inverting input of the summer amplifier 2.

14.2.2 The capacitive transducer

An elastic diaphragm made of conducting material, placed at small distance d from a flat rigid conducting electrode, is a capacitor whose value C depends on d. If we charge this capacitor with a voltage source E_o through a resistor R, as in Figure 14.13a, every displacement of the

---

[75] The SI unit (see http://en.wikipedia.org/wiki/International_System_of_Units) for pressure are pascal (Pa) and newton/square meter (N/m²). However other units as Torr (1mm Hg), atmosphere or bar (1 atm = 760 Torr = 101.32 kPa; 1 bar = 750 Torr), are still frequently used.

diaphragm induces an electric signal V(t) across the capacitor. This proves that this circuit may be used as *capacitive microphone*.

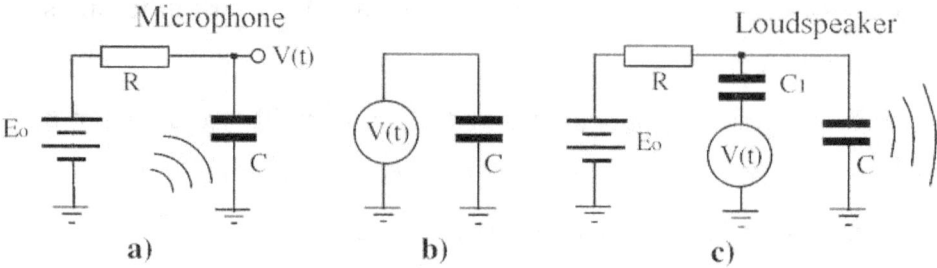

Figure 14.13

The RC low-pass filter in Figure 14.13a has output impedance $Z_C \| R$, where R includes the internal resistance of the voltage source E_o. The pressure changes associated to an acoustical wave will be faithfully transformed into a voltage signal for frequencies $\omega > 1/RC$.

The capacitive microphone is a *reversible* transducer; in Figure 14.13b the voltage source drives the capacitor with a sinusoidal signal $V(t) = V\cos\omega t$ produces an attractive force acting onto the diaphragm that is proportional to $(V\cos\omega t)^2$, so that the pressure is modulated at the frequency $2\omega t$. In order to generate an acoustical wave proportional to V(t), instead of $[V(t)]^2$, we must bias the capacitor with a d.c. voltage $E_o > V(t)$, and add the modulating voltage V(t) through a coupling capacitor $C_1 \gg C$, as in Figure 14.13c: the transfer function in this case is[76] $j\omega RC_1 / [1 + j\omega R(C + C_1)]$; the *capacitive loudspeaker* band pass is therefore $1/RC < \omega < 1/RC_1$.

14.3. Light sensors

The light flux may be defined as *"energy carried by electromagnetic waves with wavelength between 100 nm (near ultraviolet) and 10 µm (near infrared)"* (alternatively photons with energy between 12 eV and 0.12 eV). Human eye, however, is blind over a large portion of this spectral range, so that we normally consider the *visible light* that has wavelength in the range from $\lambda = 0.38\,\mu m$ to $\lambda = 0.78\,\mu m$ (figure 14.14).

The *scotopic vision curve* (eye sensitivity in darkness) is mainly due to the *rod cells receptors*, in the retina, and the *photopic curve* (eye sensitivity in well lit conditions)

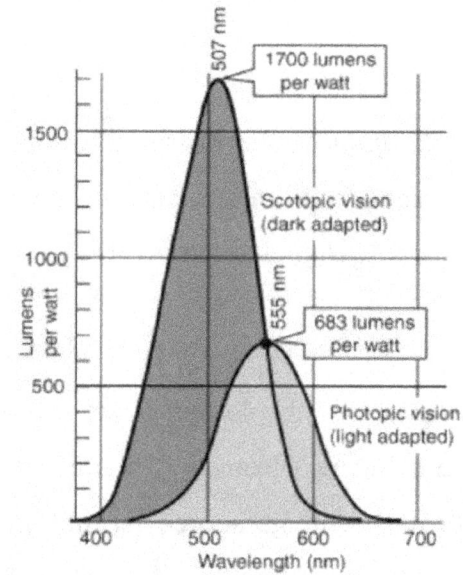

Fig. 14.14

---

[76] See chapter 5 and Appendix B.

includes the *cone cell receptors*.[77]

When the wavelength is $\lambda > 0.8\,\mu m$ radiation is named infrared (IR), or thermal radiation, when $\lambda < 0.4\,\mu m$, is named ultraviolet (UV). For λ much smaller we have X-rays then gamma-rays, and for λ much longer we have radio waves.

There are three mechanisms of light conversion into electrical signal: *thermal* (absorbed energy converted into phonons, i.e. lattice excitations, i.e. heat), *internal photoelectric* effect (electron-hole pair generation in semiconductors), and *external photoelectric* effect (electron emission by metals). We therefore distinguish among: *thermal* sensors (thermopile, pyroelectric crystals, resistive bolometers), *semiconductor* sensors (photoresistance, photovoltaic cell, photodiode, phototransistor) and i *photomultipliers*.

There are also transducers that convert electrical signal into light: *thermal* transducers (as light bulbs), *gas discharge* transducers (as arc lamps, fluorescent tubes, gas lasers), and *semiconductor* transducers (as LED and laser diodes).

14.3. Thermal light sensors

Thermal light sensor ha generally a very *flat spectral response*: constant sensitivity from IR to UV. The *thermopile* sensor is a miniaturized thermocouple made of many (up to 200) junction pairs assembled into a small device, with reference junction shaded and active junction exposed to the radiation (Figure 14.15). Commercially available thermopiles have dimensions comparable to those of a transistor in metal case, and a sensible area of the order of 1 mm². They must work with chopped light, and at low frequency (from 5 Hz to 100 Hz).

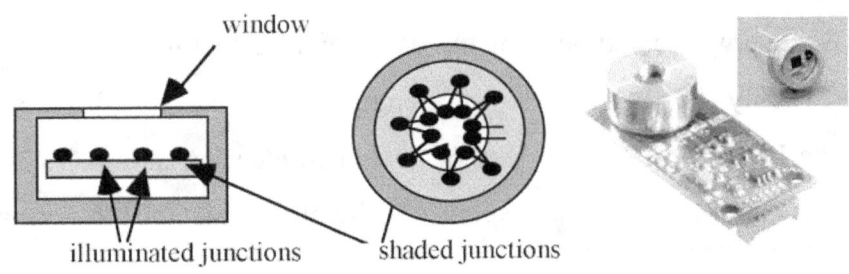

Figure 14.15

The high sensitivity types (10 V/W), have an output impedance of the order of 1 kΩ, and saturate with an input power of about 0.1 W/cm².

Pyroelectric sensors[78] exploit the property of some polar crystals to develop an electric field as response to a thermal gradient induced by heat absorption (e.g. due to electromagnetic radiation).

---

[77] The rod cells are very sensitive, but do not detect colors, while the cone cells are less sensitive but distinguish different colors.

[78] See also http://en.wikipedia.org/wiki/Pyroelectricity and http://en.wikipedia.org/wiki/Pyroelectric_crystal

Examples of pyroelectric crystals are: $PbTiO_3$, $ZrTiO_3$, $LiTaO_3$. By plating onto the faces of a pyroelectric crystal two metal electrodes we obtain a capacitor which becomes charged by the spontaneous polarization. Temperature changes produce polarization changes and therefore a weak a.c. current ($10^{-12} \div 10^{-10}$ A) across the capacitor. This current may be converted into a voltage signal by an OA with high input impedance with a high feedback resistance ($R_o \approx 2 \cdot 10^9$ Ω). Therefore a pyroelectric sensor may be seen as a current source, in parallel to a capacitor C_p (some pF) and to a resistor R_p (some 10^{12} Ω). The working frequency is in the range 10 Hz ÷ 10 kHz, with a sensitivity $\sigma_a = \partial I / \partial W \approx 1\,\mu A/W$.

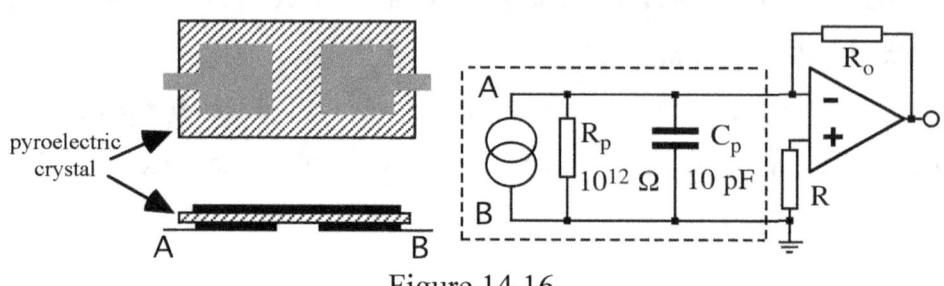

Figure 14.16

Figure 14.16 shows a possible interfacing circuit. The output signal $\delta V = R_o\, \delta I = R_o\, \sigma_a \delta W$. Due to the small sensitivity the value of R_o must be high, and therefore also the value of resistor R must be high to limit the offset V_{os} ($R = R_p \| R_o$).

14.3.2 Semiconductor light sensors

Photoresistances are made of semiconductor and exploit the *internal photoelectric effect* to convert absorbed light into *electron-hole pair generation*. Only photons with energy higher than a threshold energy E_g (Energy Gap) typical of the used semiconductor are effective, therefore the resistance decreases only for light with wavelength below the threshold wavelength λ_s[79], (generally within the IR region). The number of generated electron-hole pairs is proportional to the absorbed light flux, and the *spectral response* of the photoresistance is normally peaked at a value slightly lower than λ_s.

The photoresistance sensitivity is proportional to the lifetime τ of the charge carriers[80], and the useful frequency range is from 0 Hz to some kHz. The higher is the sensitivity the smaller is the band-pass because large values for requires longer times for the photoresistance to recovery the original value after the light pulse is finished. Photoresitors must be biased (either with d.c or a.c

---

[79] The threshold wavelength λ_s is determined by the equation $hc/\lambda = h\nu = E_g$, where h is the Plank constant, ν the light frequency, c the light speed and the energy gap E_g is the energy required to promote an electron from the valence band into the conduction band,

[80] The lifetime is inversely proportional to the crystal lattice defects and to the dopant concentration of the semiconductor.

currents). Interfacing circuits are similar to those shown in Figure 14.1 and 14.2.

The *photovoltaic cell* (also named *PhotoDiode*) is a PN junction [81], basically a diode, where the P-doped semiconductor is very thin, in order to allow incoming photons to penetrate the depletion layer where the generated electro-hole pairs may drift in the internal electric field and reach the external electrodes. This sensor does not need bias. The sensitivity has a peak close to λ_s. Response is proportional to the light intensity only for the output short-circuit current.

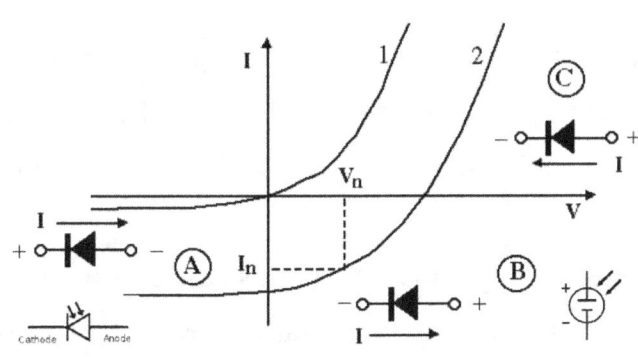

Figure 14.17

Figure 14.17 shows the characteristic I-V curves of a photovoltaic cell (where we assumed the current positive when flowing from anode to cathode): for dark condition (curve 1) and with light input (curve 2). Curve 1 is the usual diode characteristic curve, curve 2 is the same shifted downward by a quantity determined by the illumination.

In quadrant A the junction in reverse-biased and in this configuration it is normally named *photodiode* (current flowing from cathode to anode). In quadrant B the junction in forward-biased and this configuration is normally named *solar cell*: values V_n and I_n give maximum output power $W = V \cdot I$. (Solar cell I-V curves are normally traced with current positive when flowing from cathode to anode, so that the cell produces energy $W > 0$ in quadrant B, and in quadrant C it dissipates energy : $W < 0$)

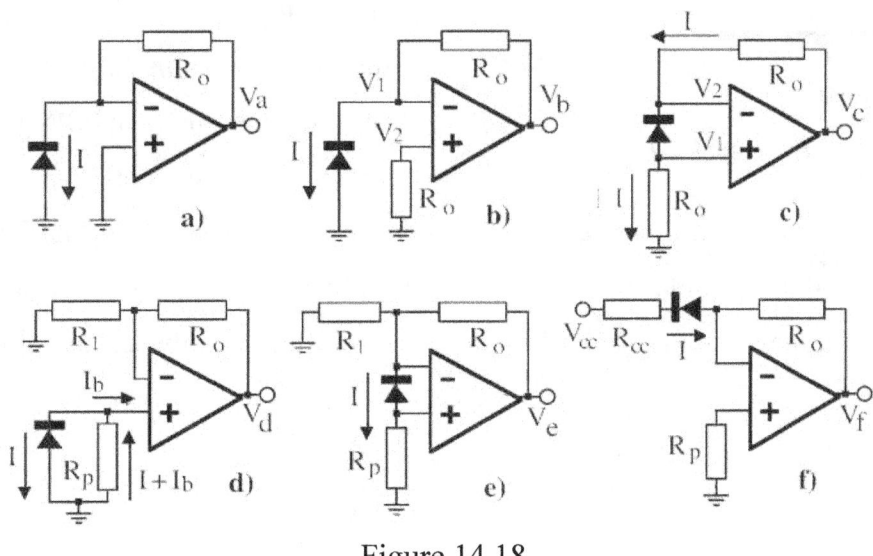

Figure 14.18

Photocells have the best linearity when short-circuited: we may get zero bias with the circuit of

---

[81] See Appendix A1.

Figure 14.18a where the current-to-voltage converter keeps the cathode at virtual ground through the feedback resistor R_o. In this circuit the output V_a is affected by the OA input bias current I_b that becomes important for small values of photocurrent I: $V_a = R_o(I+I_b)$.

We may add a balancing resistor R_o as in Figure 14.18b, so that $V_b = R_o(I+I_{os})$, reducing the error of a factor 10, but introducing a small bias to the photocell : $V_1 = V_2 = R_o I$.

The problem is completely solved by circuit of Figure 14.18c, that requires an OA with high CMRR: we get $V_1 = V_2 = R_o I$ and $V_c = R_o I + V_2 = 2 R_o I$.

An alternative circuit is shown in Figure 14.18d, where the photocurrent signal I across R_p (affected by the OA input bias current I_b), gives the output $V_d = R_p (I+I_b)(1+R_o/R_1)$, but the diode is here slightly forward biased. A zero-bias is achieved in the circuit of Figure 14.18e, that gives the output $V_e = R_p (I-I_b)(1+R_o/R_1)$.

The circuit of Figure 14.18f, provide a *reverse bias* to the photodiode. In this case an extremely small *dark current* flows across the PN junction, due the thermally-generated electron-hole pairs ($10 \, pA/mm^2$), decreasing at low temperature. This configuration, suitable for weak light fluxes, has slow response because most photons generate charge carriers out of the depletion layer that must reach the electrons by the slow *diffusion process*.

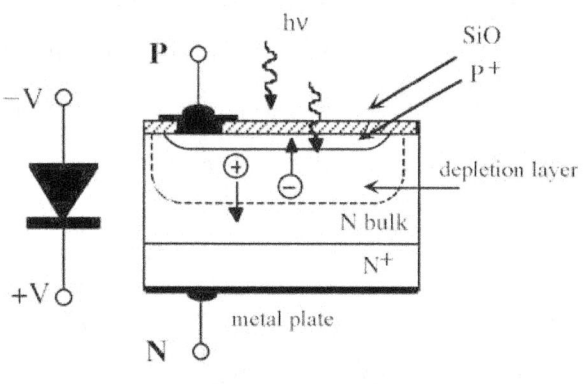

Figure 14.19

In the PIN photodiode[82], shown in Figure 14.19, the thickness of the depletion layer is increased by the reverse bias: this makes faster the response by increasing the drift velocity of the photo carriers (and by decreasing the effective capacitance).

In the *phototransistor* (NPN) the most-common variant is an NPN bipolar transistor with an exposed base region; the illuminated junction is the base-collector, which behaves as a photodiode. The inverse current is injected into the N-doped emitter region with an amplification of two order of magnitude [83]. The equivalent circuit is shown in Figure 14.20. The response is linear only for low illumination.

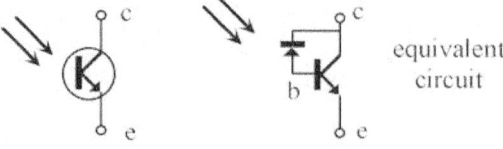

Figure 14.20

---

[82] The acronym PIN stays for P-layer/Intrinsic-layer/N-layer, because they are obtained from pure silicon wafer and pure semiconductor is named *intrinsic*.
[83] See Appendix A.

14.4. Position sensors

Position sensors may be *relative* to some reference value or *absolute*.

Absolute position sensors are, for example, those based on the time-of-flight of a traveling electromagnetic or acoustical pulse that is reflected by the target object (radar and sonar).

Relative position sensors (that measure distance changes), as the inductive sensors, need some calibration.

14.4.1. The sonar

The name SONAR is an acronym for **SO**und **N**avigation **A**nd **R**anging. This devices emits a pulse of sound and measures the time elapsed before detection of the echo produced by the pulse reflection on the target. It has many application in marine technology but also in other fields[84].

The basic structure of a sonar includes a capacitive transducer (beeper/microphone) that emits a short burst of ultrasonic pulses and detects the echo, plus a clock that measures the time elapsed .

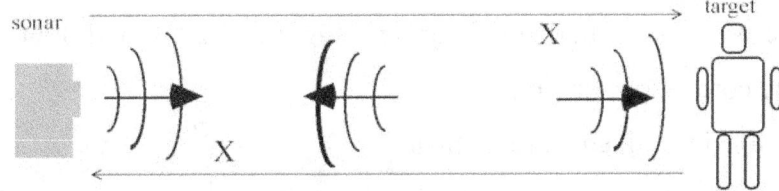

Figure 14.21

The sound velocity in air c is known: $c=(331+0.6t)$ m/s, where t is the temperature in Celsius; if T is the time elapsed, then the distance X (covered in the to/from travel) is calculated as $X=cT/2$.

The frequency of the ultrasound wave is normally some kHz, and the sensor range is typically from 0.2m to 20 m.

14.4.2. The inductive position sensors

The inductive position sensors may be LVDT (*Linear Variable Differential Transformer*): this device is a transformer the coupling between primary and secondary windings depends on the position of the mobile ferrite core. The primary coil is driven by the excitation signal and the output may be the sum or the difference of two symmetric secondary coils (Figure 14.22).

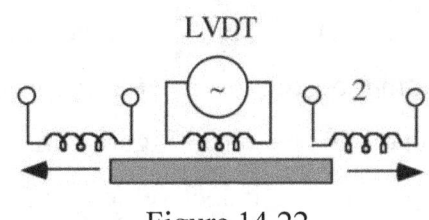

Figure 14.22

---

[84] For details on interfacing a sonar see http://www.acroname.com/robotics/info/articles/sonar/sonar.html; see also http://www.vernier.com/products/sensors/motion-detectors/. This sensor is also used in Polaroid cameras for autofocussing.

The two secondary coils may be connected in series with opposite windings: therefore the output amplitude is minimum when the core is centered and increases when it moves in both directions. Using a phase-sensitive detector (lock-in, Figure 14.23a) we may get a *linear output* proportional to the displacement (positive in one direction and negative in the opposite direction).

A simpler circuit is shown in figure 14.23b, where the output is provided by a pair of diodes and a low-pass filter.

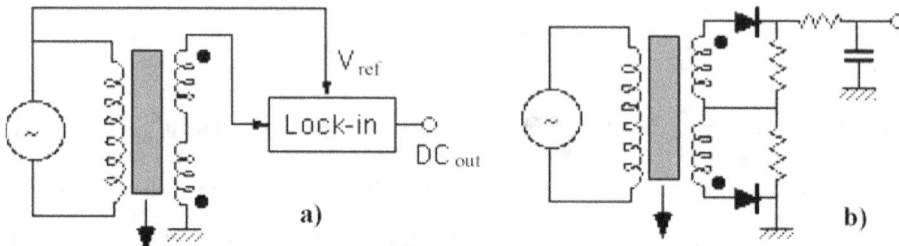

Figure 14.23

Another type of inductive position sensor is the LVRT (*Linear Variable Reluctance Transducer*) where two coils are wired as inductive half-bridge (Figure 14.24a), and the the unbalance bridge-output measures the core displacement.

Another configuration is a fixed core with a ferromagnetic object that changes the reluctance of one arm of the bridge as shown in 14.24 b..

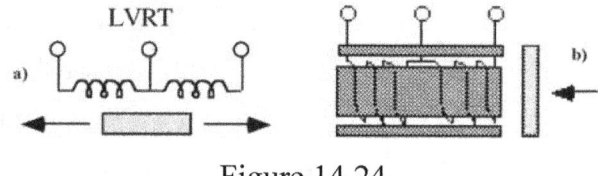

Figure 14.24

14.4.3. The resistive position sensors

The potentiometer, a three-terminal resistor with a sliding contact that forms an adjustable voltage divider, may be used as *angular position* sensor (*rotation* sensor), if the axis is mechanically linked to a rotating object, or as *linear position* sensor, depending on its geometry (Figure 14.25).

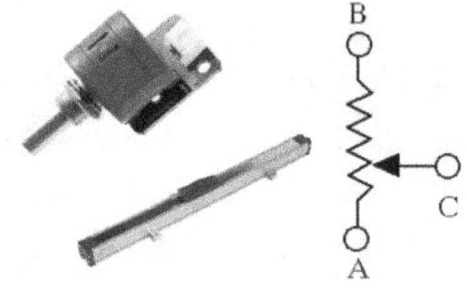

Figure 14.25

With respect to the inductive sensors, the resistive sensor have the drawbacks of friction and wearing (due to the sliding contact), but may be biased either by d.c. or a.c. voltages. The linear displacement range may be from 5 cm to 50 cm, and the angular range from 270° to 3600° (in the ten turns helipots).

14.4.3. The optical position sensors

The limited range of resistive rotation sensors is absent in the digital optical rotation sensors (*optical encoders*)[85] that use an optical threshold and a slotted disc (Figure 14.26).

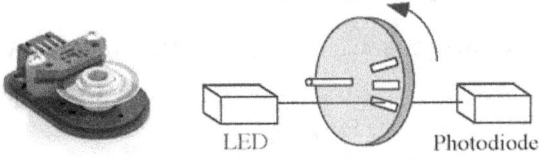

Figure 14.26

The angular position of the disc (and of the object attached to the shaft) may be accurately encoded and the resolution (defined by the angular separation of the slots) is not affected by the range. Besides the rotary models, there are commercially available also linear models, which may offer a range up to 30 m. The most popular application of such encoders is the old PC "*mouse*" (recently replaced by the "*optical mouse*" that exploits a different technique, i.e. digital image correlation).

---

[85] For a list of manufacturers see http://www.sensorsportal.com/HTML/SENSORS/RotationSens_Manuf.htm

15. The OA with double feedback

In the previous chapter we analyzed some circuits where the operational amplifier was working with double feedback (both positive and negative), and we noted that some care must be taken in order to avoid canceling the effective feedback, leading to unstable open-loop behavior. The method adopted to perform the analysis was based on the ideal OA approximation, which may lead to wrong conclusions if we do not take into account the real frequency dependence of the open-loop gain of the OA.

Let us consider the circuits shown in Figure 15.1.

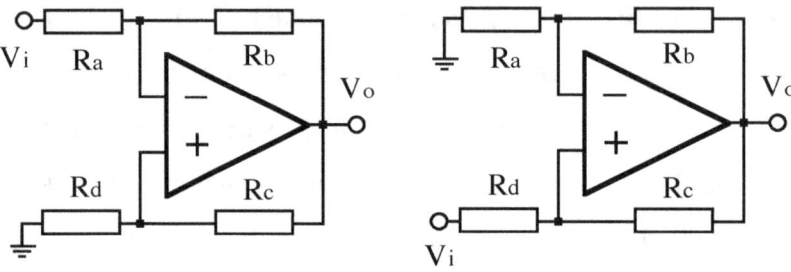

Figure 15.1

The two circuits are identical: they only differ for the choice of input, and consequently of the feedback fractions β^+ (positive feedback) and β^- (negative feedback).

The block diagram for both circuits is drawn in in Figure 15.2.

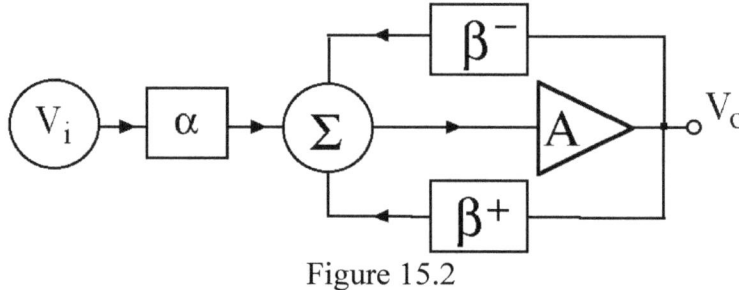

Figure 15.2

Using superposition principle we get $V_o = \alpha^{\pm} A V_i + (\beta^+ - \beta^-)V_o$, or

$$V_o/V_i = \alpha^{\pm} A/[1-\beta A], \qquad [15.1]$$

where $\beta = (\beta^+ - \beta^-)$ is the total feedback, and $\alpha^{\pm}$ is a coefficient different for the two circuits. From Figure 15.1 we get $\beta^- = R_a/(R_a+R_b)$, $\beta^+ = R_d/(R_c+R_d)$, with $\alpha^+ = R_c/(R_c+R_d)$, for the non-inverting amplifier, and $\alpha^- = R_b/(R_a+R_b)$ for the inverting amplifier.

Relation [15.1], for ideal OA ($A \to \infty$) yields

$$V_o/V_i = -\alpha^{\pm}/\beta, \text{ for } A \to \infty \qquad [15.2]$$

This is the same result obtained in chapter 3 (both in §3.1 and §3.2), without positive feedback ($\beta^+ = 0$). On the other hand, by letting $\beta^+ = \beta^-$, which yields total feedback $\beta = 0$, the OA works in effective open-loop, that predicts divergence for V_o, when $A \to \infty$.

When $\beta^+ > \beta^-$ (i.e. $\beta > 0$), relation [15.1] is cannot be *always* approximated by the simple relation [15.2] for $A \to \infty$. We must take into account the frequency dependence of both $A(s)$ and $\beta(s)$, where s is the complex frequency A general discussion of this situation is not trivial [86]. Here we will analyze some important specific cases.

If we assume the typical frequency dependence for the open-loop gain $A(s) = A_o/(1 + s/\omega_o)$ of the OA, and a real feedback β, relation [15.1] becomes:

$$V_o/V_i = T(s) = \alpha^{\pm} A_o/(1 - \beta A_o + s/\omega_o), \qquad [15.3]$$

a function with a pole on the real axis $s = \omega_o (\beta A_o - 1)$. The Laplace transform analysis (see Appendix B.6) predicts divergence when the transfer function has a pole on the real axis. Therefore the approximation [15.2] is limited by the condition $\beta A_o < 1$, for any value of A_o, also for $A_o \to \infty$.

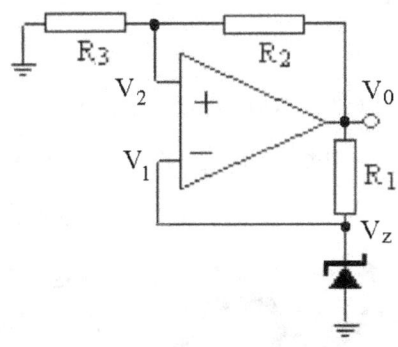

Let us consider some practical example: in §5.1 we introduced a positive feedback in the circuit of Figure 5.3, to improve a zener-stabilized voltage source. In Figure 15.3 is drawn the same circuit, but with exchanged inputs in the OA (and omitting the voltage divider for simplicity: we will assume a reverse biases zener).

Figure 15.3

The analysis is the same. By using the superposition principle: $V_o = -AV_z + A[R_3/(R_3+ R_2)]V_o$, , or $V_o[1-(R_3+ R_2)/AR_3] = -V_z(1+R_2/R_3)$, that for $A \to \infty$, gives $V_o = -V_z (1+R_2/R_3)$.

This conclusion is wrong !! In fact it predicts a forward biased zener ($V_o < 0$) contradicting our initial assumption.

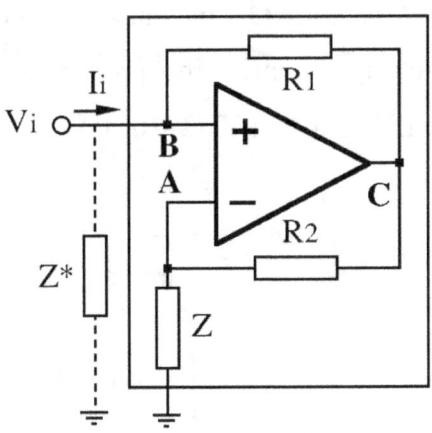

A second example of double feedback was given in §8.8: here it is easy to show that the two inputs may be exchanged as in the circuit of Figure 15.4. The performance is the same as reported by various authors[87].

Figure 15.4

---

[86] A detailed analysis may be found in *Feedback and control system analysis and synthesis*, J.D'Azzo et al.
[87] See *The Art of Electronics*, P. Horowitz et al., fig. 4.4, page 151.

16. Guide to experiments

This chapter suggests some practical exercises with the circuits described in previous chapters, giving in most cases only suitable values for the passive elements and sometimes also hints for performing elementary measurements.

16.1. Some preliminary suggestion

The simplest method to test a circuit is to mount it onto a *solderless breadboard*[88], (see Figure 16.1 and Appendix C.6) that allows fast checking without soldering the components. More complex circuits should be soldered onto a *stripboard*[89]: soldered contacts are in fact more reliable than pressure contacts.

For the IC components, however, it is better to use pressure sockets that allows avoiding overheating the IC pins with the soldering iron. Pressure sockets (Figure 16.2) should be soldered to the stripboard before inserting the IC.

Figure 16.2

Figure 16.1

Ancillary basic instrumentation is: dual power supply (±15 V, possibly with adjustable outputs), digital tester (2 or 4 digits), a signal generator (1 Hz÷100 kHz) and an oscilloscope (2 channels).

The default OA is a generic one (μA741 or equivalent), the default bias voltage is dual (V_{cc} = ±15 V), filtered by two capacitors connected to common ground. Generic OA may also be used as comparators, but a better choice is to use models that avoid latch-up (specified in Appendix D.3). For the *timer 555* the timing RC filter should use values in the ranges 10 kΩ÷10MΩ and 100 pF÷10 μF. The minimum pulse width is about 10 μs.

Signal generators: the output impedance of commercial oscillators is normally 50 Ω; many models offer also an adjustable d.c. offset(that might be useful for exercises with circuits of chapter 8. The suggested amplitude is 1 V peak-to peak.

A fast test of *phase relation* between two signals is achievable using the oscilloscope in X-Y mode: this means that one signal is fed to vertical deflection amplifier (channel Y) and the other signal to the

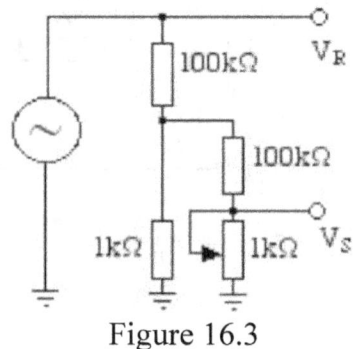

Figure 16.3

---

[88] See http://en.wikipedia.org/wiki/Breadboard
[89] See http://en.wikipedia.org/wiki/Stripboard

horizontal deflection amplifier (channel X).

The Lissajous curve traced onto the oscilloscope screen in X-Y mode (Figure 16.4) by the synchronous signals V_x and V_y, gives the phase ϕ delay. Assuming equal amplitude: $V_x = V_y = V_o$, (eventually by adjusting the channel's gain) we get: $V_x = V_o \cos(\omega t)$ and $V_y = V_o \cos(\omega t + \phi)$. If $\phi = 0$ (or $\phi = \pi$), the trace is straight line with slope $\pi/4$ (or $3\pi/4$) with respect to X axis, and if $\phi = \pi/2$ (or $\phi = 3\pi/2$), the trace is a circle. For intermediate values of the trace ϕ is an ellipse, whose major axis is in the first quadrant for $0 < \phi < \pi/2$ and in the third quadrant for $\pi/2 < \phi < \pi$. In this case, after centering the ellipse on the screen, we measure the intercept $Y(0)$ of the trace on the Y axis and we get $\phi = \arcsin[Y(0)/V_o]$, as proven by the relations $V_x/V_o = x = \cos \omega t$, and $V_y/V_o = y = \cos(\omega t + \phi) = \cos \omega t \cos \phi - \sin \omega t \sin \phi$.

For $x = 0$, $\sin \omega t = 1$ and therefore $y(0) = \sin \phi$.

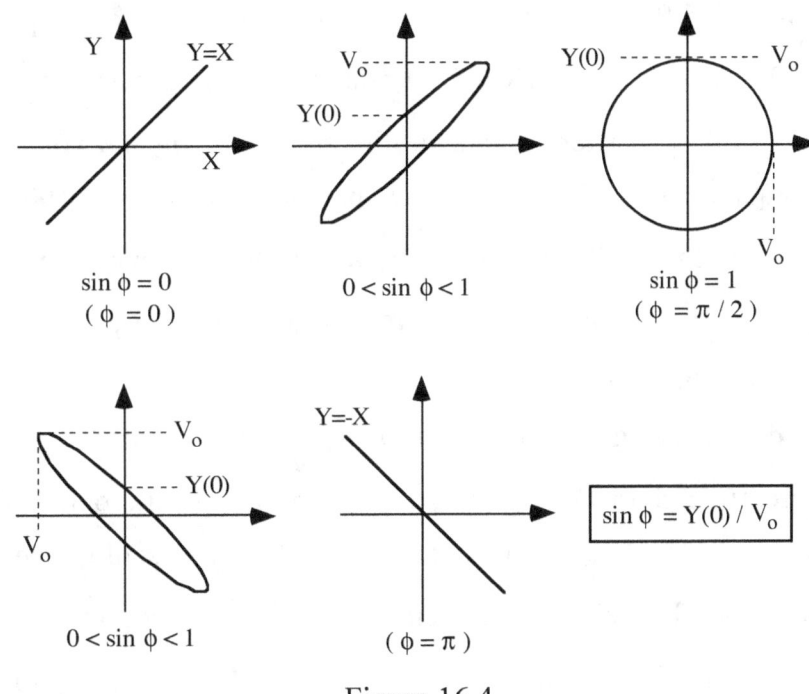

Figure 16.4

16.2 Exercises

In this section each exercise refers to the circuit shown in the corresponding Figure

Figure 3.1, 3.2 and 3.3 Inverting and non-inverting amplifier
Choose $R_i = 1$ kΩ, $R_o = 1 \div 10$ kΩ, $R = 0$ Ω, $V_i = 0 \div \pm 10$V. Measure V_o for several V_i values and for different R_o values within the suggested ranges. Note how changes the input voltage range for linear behavior with different gain values. In the circuit of Figure 3.2 set $R_o = 0$ Ω, and verify that you get a *follower* (Figure 3.3).

Figure 3.4 Differential amplifier

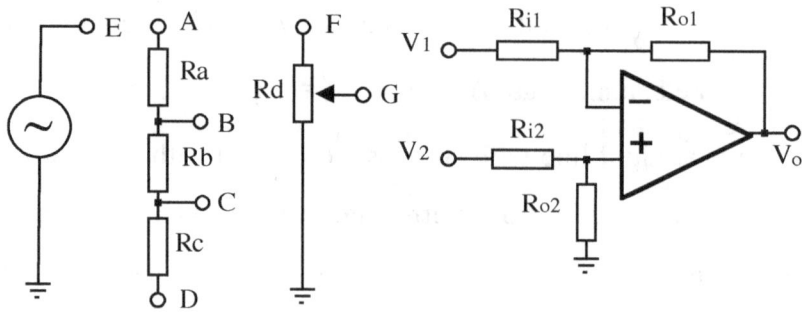

Figure 16.5

a) Offset zeroing, with G=100. (we neglect I_{os}). $R_{i1}=R_{i2}=1\,k\Omega$, $R_{o1}=R_{o2}=100\,k\Omega$. Short the inputs V_1 and V_2 to ground. Achieve offset-null ($V_o=0$) through a 20 kΩ potentiometer connected to pins 1–5 (see § D.2.1 or § D.2.2).

b) Measurement of differential voltage with G=10. Choose: $R_{i1}=R_{i2}=10\,k\Omega$ and $R_{o1}=R_{o2}=100\,k\Omega$, (carefully select 1% resistors, or use a small resistance trimmer in series to the smallest resistor to balance the circuit. Using the voltage divider ABCD shown in Figure 16.1 ($R_a=R_c=10\,k\Omega$ potentiometers, $R_b=100\,\Omega$ potentiometer, $V_A=+15V$, $V_D=-15\,V$). Connect B to V_1 and C to V_2 ; measure with a multimeter voltages V_C, V_B, V_C-V_B and V_o, for several values of R_a, R_b, R_c. Exchange inputs V_1 and V_2, and repeat the measurements.

c) Measurement of differential gain, with G=100 ($R_{i1}=R_{i2}=1\,k\Omega$ and $R_a=R_c=10\,k\Omega$). With reference to Figure 16.5: short D to ground and connect A to the output E of a sinusoidal oscillator (frequency $\approx$ 1kHz, $V_{pp}\approx 1V$) and to channel-1 of the oscilloscope; connect B to V_1 and C to V_2, and V_o to channel-2 of the oscilloscope. Measure V_o and V_E and calculate the differential gain $A_d=V_o/(V_C-V_B)$, where $(V_C-V_B)\approx -V_E R_b/(R_a+R_b+R_c)$, for different values of R_b.

d) Measurement of common-mode gain. ($R_d=10\,k\Omega$). Connect E to F and G to both V_1 and V_2. Connect V_G to channel-1 and V_o to channel-2. Note that $V_{cm}=V_G$, $A_d=R_o/R_i$ and $V_o=A_d(V_{os}+V_d)+A_{cm}V_{cm}$. Remember that V_{os} is a d.c. voltage while V_G is a.c. voltage. Letting $V_G=0$ (you get $V_{os}=V_o R_i/R_o$) adjust the offset-null. Now you may measure $A_{cm}=(V_o/V_G)$ for several values of V_G. Minimize A_{cm} by adjusting resistances $R_{o1}, R_{o2}, R_{i1}, R_{i2}$

Figure 3.8 Evaluate V_{os}, I_{b2} and I_{os}

a) Choose $R_i = R = 100\,\Omega$, $R_o = 10\,k\Omega$. By neglecting $\Delta V_o = -R_o I_{os}$ we get $V_{os} \approx V_o R_i/R_o$. Measure V_{os} for different OA (e.g. µA741 and TL081). Adjust the offset-null with a 20 kΩ trimmer connected to pins 1-5 (§D.2.1 or §D.2.). Follow the offset drift while heating the OA with an hot soldering iron (do not exceed in heating).

b) $R = 1\,M\Omega$ pot. Remove R_i (replace R_o with a short). You get $V_o = V_1 = V_2 + V_{os} = -RI_{b2}$ Assuming $V_{os} \approx 0$ (previously adjusted to zero) you get: $I_{b2} = -V_o/R$. Note that the output offset depends on R Compare I_{b1} values for different OA.

c) Choose $R_i = R = 10\,k\Omega$, $R_o = 1\,M\Omega$. Because $R \approx R_i \| R_o$, we get $V_o = V_{os} - R_o I_{os} \approx -R_o I_{os}$, that gives an evaluation of I_{os}. By removing R you get $V_o = R_o I_{b1}$, i.e. an evaluation of $I_{b1} = V_o/R_o$.

Figure 4.1 Differential amplifier with variable gain
Choose $R_i = R_o = R_a = 100\,k\Omega$ (both branch' and branch"), $R = 100\,k\Omega$ pot in series to 1 kΩ (0.01 < x < 1.01). Measure the differential gain while changing R. Input a.c. signal as in Figure 16.1 ($R_a = R_c = 10\,k\Omega$, $R_b = 100\,\Omega$).

Figure 4.2 Differential amplifier with linearly variable gain
OA = TL082, $R_1 = 10\,k\Omega$, $R_o = 100\,k\Omega$, R = 100 kΩ pot, R' = 100 kΩ. Demonstrate that the differential gain is: $G = R_o/R_i[(x+R'/R)/(1+R'/R)]$, 5 < G < 10. Note that, with the smallest (R'+xR) value, the output Vo range is reduced.

Figure 4.3 Differential amplifier with linearly variable gain
OA = TL082, $R_1 = 10\,k\Omega$, $R_o = R = 100\,k\Omega$, xR = 100 kΩ pot, R' = 100 kΩ. Demonstrate that the differential gain is: G $R_o/R_i(x+R'/R)$, 10 < G < 20.

Figure 4.4 Differential amplifier with linearly variable gain
Choose OA = TL082, $R_2 = R'_2 = R_1 = R'_1 = 10\,k\Omega$, xR = 100 kΩ pot. Note that the gain does not depend on R_2 and R'_2, until $R_2 = R'_2$.

Figure 4.5
$R_o = R'_o = R_1 = R'_1 = 10\,k\Omega$, xR = 100 kΩ pot in series to 10 kΩ. 2 < G < 40.

Figure 4.6 Instrumentation amplifier
OA = TL084. $R_o = R_1 = R_2 = R_3 = 100\,k\Omega$, xR = 100 kΩ pot in series to 10 kΩ. Note that R_2 may be different from R_3: the gain value is determined by their sum.

Figure 4.7 Amplifier with positive or negative gain

Values suggested for ($-1 < G < +1$): $R_o = R_1 = 10\,k\Omega$, $R = 100\,k\Omega$ pot, $R_2 = \infty$ (removed). Values suggested for ($-10 < G < +10$): $R_o = R_1 = 1\,k\Omega$, $R = 10\,k\Omega$ pot, $R_2 = 1.11\,k\Omega$.

Figure 5.2 Voltage source

$V_{cc} = +15V\,/\,0V$. $R_2 = R_1 = 10\,k\Omega$. R_L 10 $k\Omega$ in series to an amperometer (multimeter). The voltage V, may be obtained from $+V_{cc}$ through a voltage divider as V_G in Figure 16.5.

Figure 5.3

Use 6.9 V zener (e.g. LM329) with $R_b \approx 5\,k\Omega$, $R_a \approx 10\,k\Omega$. For $V_o = +10V$, choose $R_1 = 3.3\,k\Omega$ ($I_z \approx 1mA$), $R_2 = 10\,k\Omega$, $R_3 = 22\,k\Omega$. For $V_o = -10V$, reverse the zener and the diode.

Figure 5.4

As the previous circuit, but exchange the values of R_2 and R_3.

Figure 5.5 Twin voltage source

Use 6.9 V zener, $R_1 = 2\,k\Omega$, $R_2 = 10\,k\Omega$, $R_3 = 50\,k\Omega$. Using power OA (µA759, µA791, TC365, L165, 3571), this may give a dual power supply $\pm V_z$, from a single one with $V_{cc} > 2V_z$.

Figure 6.1 , 6.2 , 6.3 Current sources

$R_i = 1\,k\Omega$, $R_L = 10\,k\Omega$ pot in series to an amperometer. Measure I_L, for various R_L values, as a function of V_i, between 0 and $V_{cc}/2$ for circuit 6.1a, and from 0 and $-V_{cc}/2$ for circuit 6.1b. Check the ranges of I_L and of R_L within which the circuit does work properly. In the circuits of Figures 6.2 , 6.3 use a battery for V_i, or another voltage source referred to a ground insulated from the ground of the power supply used to bias the OA.

Figure 6.4 Voltage controlled current source

$R_1 = R_2 = 10\,k\Omega$, $R_o = R_3 = 1\,k\Omega$, $R_L = 5\,k\Omega$ pot in series to an amperometer. The capacitor (some pF) may be be placed in parallel to R_o. Verify that I_L does not depend on R_L, and that is may be controlled by the input voltage V_i (that must be generated by a low output impedance source, to avoid affecting the R_1 effective value.

Figure 6.5

$R_1 = 1\,k\Omega$, $R_2 = 9\,k\Omega$, $R = 10\,k\Omega$, $C = 10$ nF, $R_L = 5\,k\Omega$ pot in series to an amperometer. Verify that I_L does not depend on R_L, and that is may be controlled by the input voltage.

Figure 6.6
OA = TL082, $R_1 = R_2 = 10\,k\Omega$, $R_3 = R_4 = 1\,k\Omega$, $R_5 = 1\,k\Omega$ in series to a $100\,k\Omega$ pot. $R_L = 5\,k\Omega$ pot in series to an amperometer. C = some nF. Verify that I_L does not depend on R_L, and that is may be controlled by R_5 or V_i.

Figure 6.7
OA = LF356, 1.2 V zener, $R_o = 5\,k\Omega$, $R = 1\,k\Omega$ in series to a $100\,k\Omega$ pot, $R_L = 10\,k\Omega$ pot in series to an amperometer. Measure I_L as a function of R and R_L; evaluate I_{Lmax} and R_{min}.

Figure 7.1 Half-wave rectifier
OA = μA741, diodes = 1N914, $R_o = R_L = 1\,k\Omega$. Small input signal ($|V_{in}| < 1\,V$). Use the oscilloscope in X–Y mode. Compare results for circuits of Figures 7.1a and 7.1b, with and without diode D2. Replace bipolar OA with FET-input OA, choose $R_o = 100\,k\Omega$, diodes 1N456÷1N459, and note the different behavior.

Figure 7.2 Inverting half-wave rectifier
$R = R' = R_L = 10\,k\Omega$, FET input OA, or, using bipolar OA, insert a resistor (R/2) at the non-inverting input. Use the oscilloscope in X–Y mode.

Figure 7.3 Full-wave rectifier
$R_o = R_1 = R'_1 = R_2 = R'_2 = 10\,k\Omega$, $R_i = 10\,k\Omega$ pot in series to $1\,k\Omega$, for gain trimming. $R'_o = 2\,k\Omega$ pot in series to $9\,k\Omega$, for output symmetry.

Figure 7.4
$R_o = 10\,k\Omega$ pot in series to $10\,k\Omega$, $R' = R = 10\,k\Omega$. Load the output with $10\,k\Omega$ resistor. Use C≈100 pF capacitance to avoid oscillations. Note that the value of R_o does not affect the circuit's behavior.

Figure 7.5
$R_1 = R'_1 = R_i = 2R_2 = 10\,k\Omega$ (R_2 may be the parallel of two $10\,k\Omega$ resistors), $R_o = 50\,k\Omega$ pot in series to $10\,k\Omega$ for gain trimming.

Figure 7.6
$R' = R = 10\,k\Omega$, $R_1 = 5\,k\Omega$. Use C≈10 pF in parallel to D2, or in parallel to both diodes to prevent oscillations.

Figure 7.7
Must be G=R'/R. Choose G=2, i.e. $R_2 = 3R$; e.g. $R_2 = 3 k\Omega$, $R' = 2 k\Omega$, $R_1 = R = 1 k\Omega$.

Figure 7.8
$R_1 = R_2 = R_3 = 10 k\Omega$. $R_4 = 20 k\Omega$ (R_4 may be the parallel of two 10 kΩ resistors). A capacitance in parallel to D1 helps avoiding oscillations.

Figure 7.9 Peak detector
$R \approx 100 k\Omega$. $C \approx 10 nF$ (ceramic, must be a small value if you need a fast peak-detector). Choose diodes with low leakage (e.g. 1N458). OA : FET-input. Use a double-channel digital scope to compare input and output while manually changing input (starting from Vi=0) with a potentiometric divider as that shown in Figure 16.5.

Figure 7.10 Improved peak detector
$R_1 \approx 100 k\Omega$, $R_2 \approx 20 k\Omega$. $C \approx 10 nF$. OA1 = generic (e.g. µA741), OA2 = FET-input OA (e.g. LF356). Or use dual FET-OA (e.g. LF353, TL082).

For all the exercises of Chapter 8 use a sinusoidal oscillator (e.g. 1V amplitude) and a two-channel oscilloscope to observe V_o and V_i while changing the frequency.

Figure 8.1b Integrator
$R = 10 k\Omega$, $R_o = 1 M\Omega$, $C = 10 nF$. Use a sinusoidal oscillator without offset (it would be amplified of a factor 100!). Start from high frequency ($\approx 50 kHz$) decreasing until $|V_o| = |V_i|$. Evaluate the phase-lag of V_o with respect to V_i. Draw the plot of V_i/V_o versus ω, and verify that the slope is RC. To remove possible residual input offset, use an high-pass filter between oscillator and integrator input. Switch to a square-wave oscillator to drive the input and observe that the output is a triangle-wave.

Figure 8.2b Differentiator
$R_i = 1 k\Omega$, $R = 100 k\Omega$, $C = 10 nF$. Start from low frequency ($\approx 100 Hz$) increasing until $|V_o| = |V_i|$. Evaluate the phase relation between V_o and V_i. Switch to a square-wave oscillator to drive the input and observe the output signal.

Figure 8.4 Active low-pass filter
Choose $R_1 = R_3 = R_4 = 10 k\Omega$, $C_2 = C_5 = 10 nF$. You get G=1, $f = \omega_0/2\pi \approx 1.59 kHz$ and $\zeta = 1.5$. If you double the values of resistors you see that the values of G and ζ do not change, and ω_0 is

reduced of a factor 2. Increase C_2 of a factor 10 and decrease C_5 of a factor 10, you see that G and ω_0 do not change, and ζ scales of a factor 10. Observe that the maximum of the transfer function amplitude is approx. $G/2\zeta$ (when $\zeta \ll 1$). Replace R_1 with a 100 kΩ pot and observe the dependence of G and ζ on R_1.

Figure 8.5 Active high-pass filter
Choose $C_1 = C_3 = C_4 = 10$ nF, $R_2 = R_5 = 10$ kΩ. You get G = 1, $f_0 = \omega_0/2\pi \approx 1.59$ kHz and $\zeta = 1.5$. If you double the values of capacitances you see that the values of G and ζ do not change, and ω_0 is reduced of a factor 2. Increase R_5 of a factor 10 and decrease R_2 of a factor 10, you see that G and ω_0 do not change, and ζ scales of a factor 10. Observe that the maximum of the transfer function amplitude is approx. $G/2\zeta$ (when $\zeta \ll 1$).

Figure 8.6 Band-pass filter
Choose $R_1 = R_2 = 10$ kΩ, $R_5 = 20$ kΩ, $C_3 = C_4 = 10$ nF. You get G = 1, $f_0 = \omega_0/2\pi \approx 1.59$ kHz and Q = 1. If you double the values of capacitances you see that the values of G and Q do not change, and ω_0 is reduced of a factor 2. Divide R_1 and R_2 by 2, and double R_5, you'll se that ω_0 does not change, while Q doubles and G is multiplied by 4.

Figure 8.12 Low-pass VCVS
Choose $R_1 = R_2 = 10$ kΩ, $C_3 = C_4 = 10$ nF. You get $f_0 = \omega_0/2\pi \approx 1.59$ kHz and $\zeta = 1$. Increase C_3 of a factor 10 and decrease C_4 of a factor 10, you see that ω_0 does not change, while ζ scales of a factor 10.

Figure 8.13 High-pass VCVS
Choose $R_3 = R_4 = 10$ kΩ, $C_1 = C_2 = 10$ nF. You get $f_0 = \omega_0/2\pi \approx 1.59$ kHz and $\zeta = 1$. Multiply R_4 by 10 and divide R_3 by 10, you see that ω_0 does not change, while ζ scales of a factor 10..

Figure 8.15 State-variable filter
With $R_1 = R_2 = R = 10$ kΩ, $C = 10$ nF, you get $f_0 = \omega_0/2\pi \approx 1.59$ kHz, $\zeta = 0.5$, $Q = G_1 = G_2 = 1$. Compare the three output signals with the input while changing frequency. Letting $R_2 > 100$ kΩ you $\zeta \approx 0.1$, $G_1 \approx 2$, $Q \approx 5$.

Figure 8.16 Notch filter
Choose $R_1 = R_2 = 10$ kΩ, $C = 10$ nF and measure ω_0 and Q.

Figure 8.18 Band-pass NIC filter
With $R_a = R_b = 10\,k\Omega$, $C_a = C_b = 10\,nF$, $R_2 = 10\,k\Omega$, $R_1 = 15\,k\Omega$ in series to a $5\,k\Omega$ pot, you get $f_0 = \omega_0/2\pi \approx 1.59\,kHz$. By trimming the pot, adjust the value $G^* = R_1/R_2$ approaching the value 2, and observe the Q divergence.

Figure 8.19 , 8.20 Gyrator
Choose $R_1 = R_2 = R_3 = R_5 = 10\,k\Omega$, $C = 100\,nF$. You get $L^* = 10^8\,C$ = 1 mH. Make an R*L* low-pass filter by driving the effective inductance L* through a resistance $R = 10\,k\Omega$ (Figure 16.6). Measure the break frequency of the R*L* filter. Note that the effective R* resistance is $R + R_s$, where R_s is the output impedance of the oscillator (typically 50Ω). To avoid oscillations use a small capacitance (≈10 pF) in parallel to R_3, in circuit of figure 8.19, or R_5, for figure 8.20.

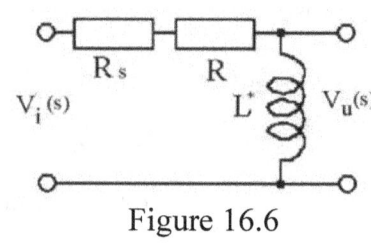
Figure 16.6

Figure 8.21 Capacitance multiplier
Use $R_i = 10\,k\Omega$, $R_o = 100\,k\Omega$ or $1\,M\Omega$, $C = 10\,nF$. You get $C^* = 11\,C$ or $101\,C$. Measure the break frequency of the low-pass R*C* filter, as above. Here the damping capacitor (≈47 pF) should be placed in parallel to R_o.

Figure 8.22 Capacitance multiplier
$R_1 = 1\,k\Omega$, $R_2 = 100\,k\Omega$, $C = 1\,nF$. These values give an equivalent capacitance is $C^* \approx 100\,nF$ in series to $R_p = R_1 \| R_2 \approx 1\,k\Omega$. Measure the break frequency of the low-pass R*C* (with R*=R+R_s). The transfer function of this filter is $[(1+sR_pC^*)/(1+s\{R_p+R^*\}C^*)]$.

Figures 8.24 , 8.25   IC Active filters
Beging with $R_{F1} = 10\,k\Omega$, $R_{F2} = k\Omega$, $R_i = 10\,k\Omega$, $R_Q = 1\,k\Omega$. (for fig. 8.25 : $R_{F3} = R_{F4} = R_{F5} = 10\,k\Omega$) Then observe the changes in G, Q and ω_0, by changing the starting values.

Figure 9.1 Comparators
Drive the OA with a sinusoidal oscillator, $f \approx 1\,kHz$, $V_{pp} \approx 5\,V$, as in Figure 16.7. Use the oscilloscope in X_Y mode to observe how changes the transfer function while varying the reference voltage V_R.

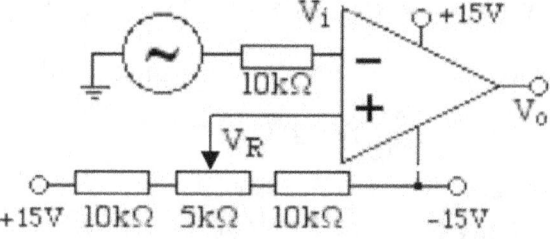
Figure 16.7

Figure 9.2 e 9.3   Non-inverting and inverting comparator

Drive the OA with a sinusoidal oscillator, $f \approx 1\,kHz$, $V_{pp} \approx 5\,V$, as in Figure 16.8. ($R_1 = R_2 = R_3 = 10\,k\Omega$). observe how changes the transfer function while varying the reference voltage V_R. Replace R_1 with a 5 kΩ pot and observe how changes the hysteresis with R_1 values. Connect V_i to the voltage divider (SW$_1$) and measure the threshold voltage, for different values of V_R and R_1. Interchange V_R and V_i (circuit of Figure 9.3) and repeat the measurements.

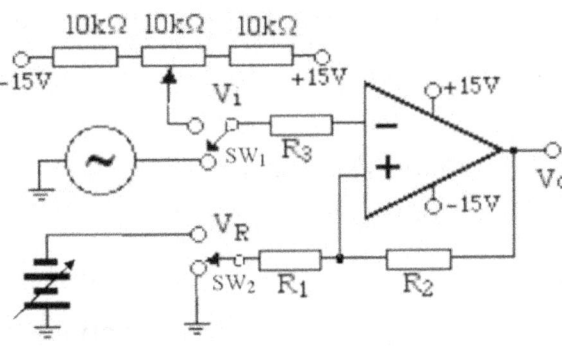

Figure 16.8

Figure 9.4   Bipolar multivibrator

$C = 10\,nF$, $R_1 = R_2 = 100\,k\Omega$, $R = 100\,k\Omega$ pot. Connect V_o to channel-1 and V_1 to channel-2. Measure the square-wave period as a function of R. Swap R with R_2 and measure the square-wave period as a function of R_2 (pot).

Figure 9.5

$C = 10 \div 100\,nF$, $R_1 = R_2 = 10\,k\Omega$, $R_o = 1\,k\Omega$. $R' = R'' = 100\,k\Omega$ pots in series to 100 Ω. Generic diodes and 5.8V zener. Measure the period and pulse width as functions of R' and R". Measure amplitudes of V'_o and V_o for various bias voltages (Vcc).

Figure 9.6   Unipolar multivibrator

$C = 100\,nF$, $R_1 = R_2 = 100\,k\Omega$. $R_3 = R = 100\,k\Omega$ pot in series to 10 kΩ. $\pm V_{cc} = +15\,V$, 0 V. Connect V_o to channel-1 and V_1 to channel-2 and measure the frequency as function of R and R_3. Repeat with $C = 10\,nF$. Swap R_2 and R_3 and measure the pulse width as function of R_2 (pot). Repeat measurements with R_2 and $-Vcc$ connected to $-15V$.

Figure 10.2 and 10.3   Wien-bridge oscillator

Figure 10.3a: $R_2 = R_3 = 15\,k\Omega$, $C_2 = C_3 = 10 \div 100\,nF$, $R_o = 200\,\Omega$ pot in series to 200 Ω. The value of R_o must be adjusted in order to stabilize the amplitude and minimize the sine wave distortion. Use filament lamp 12V-20 mA or da 24V-50 mA, (with resistance at 10 mA of about 500 Ω and 250 Ω, respectively). If R_L is the value at the working current, the R_o value should be adjusted to about 2 R_L. Figure 10.3b: The NTC thermistor may be 4.7 kΩ Philips (mod. 232262721472), with $R_1 = 5\,k\Omega$ pot.

Figure 10.4
R = 20 kΩ pot, R_1 = 20 kΩ, R_o = 47 kΩ, R_f = 10 kΩ. Adjust R for best stability and minimum distortion.

Figure 10.5 Phase shifter
R_o = 10 kΩ, C = 100 nF, R = 100 kΩ pot. Oscillator : $5V_{pp}$ (may be one of the Wien-bridge circuits previously tested). Use X-Y mode oscilloscope to evaluate the phase shift versus frequency, for both circuits 10.5a) and 10.5b), and for various R,C values. Note that amplitude does not depend on ω.

Figure 10.7 Double phase shifter oscillator
OA = TL084, R_o = R = 15 kΩ, Z_1 = Z'_1 = C = 100 nF, Z_2 = Z'_2 = 15 kΩ, R'_o = 10 kΩ in series to a 10 kΩ pot for amplitude adjustment. Stability is improved by adding two diodes and a small resistance (200 Ω). in parallel to the R'_o feedback resistor (as in Figure 10.4). Frequency may be changed by varying Z_2 and Z'_2. High frequency self-oscillations may be damped with a small capacitor (e.g. 22 pF), in parallel to R'_o.

Figure 10.8 Quadrature shifter
R_1 = R_2 = 10 kΩ, C_1 = C_2 = 10 nF. The oscilloscope in X-Y mode will give an ellipse with orthogonal axes: changing frequency the π/2 phase shift will not change.

Figure 10.9 Quadrature oscillator
OA = TL082, R_1 = R = 15 kΩ, R' = 10 kΩ in series to a 10 kΩ pot, C = C' = C_1 = 10 nF. In parallel to C_1 : 200 kΩ pot , in series to 200 kΩ in parallel to two diodes. Begin with R' slightly smaller than R to trigger oscillation, then adjust it to minimize distortion.

Figure 10.10 Quadrature oscillator
R = R' = 10 kΩ, C = C' = C'' = 100 nF, R_o = 100 kΩ in series to a 20 kΩ pot, R_f = 10 kΩ. Begin with maximum R_o value, then reduce it to minimize distortion.

Figure 10.11 Square/triangular wave generator
OA = TL082, R_1 = 1 kΩ, R_2 = 3.3 kΩ, R = 10 kΩ, C = 100 nF. By replacing R with two different resistances each in series with a diode, (with different polarities) the integrator current is different for the rising and falling slopes in the triangular wave, so that the output comparator gives pulses with different width.

Figure 10.12 Square/triangular wave generator
OA = TL082, R = 10 kΩ, R_1 = 3.3 kΩ, R_2 = 1 kΩ, R_G = R_F = 5 kΩ pots, C = 100 nF, R_T = R_Q = symmetric voltage dividers with two 10 kΩ resistors in series to 1 kΩ pot, 6.9 V zener. Vcc = ±15 V. Adjust the symmetry of triangular and square waves (R_T and R_Q values). Adjust the amplitude of V_T (by trimming R_G), then restore the same frequency value by trimming R_F.

Figure 10.13
OA = TL082, R = R_2 = 10 kΩ, C = 10÷100 nF, R_1 = 10 kΩ pot. Change amplitude (and therefore frequency) of V_T by varying R_1. Insert a resistor (≈1 kΩ) between the OA1 output and V_Q output, and a twin zener between the output V_Q and ground to stabilize the amplitude.

Figure 10.14 Quadrature square/triangular wave generator
OA = TL084, R = 10 kΩ, C = 10 nF. A four-channel oscilloscope may make easier to compare the four outputs. Try changing one R (or C), and observe the effect on the signals.

Figure 10.15 Voltage to frequency converter
OA = TL082, R = 100 kΩ, R_1 = R_2 = 10 kΩ, R_3 = 1 kΩ, C = 1 nF (o 10 nF), D = 1N914, V_{cc} = ±15 V. Measure the frequency as a function of the (positive) input voltage. Reverse diode and use negative input voltage and observe the result.

Figure 10.16 Frequency-to-voltage converter
OA = TL082, R_1 = 1 kΩ, twin Zener 6.9V, R = 1 MΩ, C = 1 nF, C_1 = 1 μF, diodes 1N914. Measure the input voltage as a function of the input frequency. Evaluate the minimum frequency that warrants linearity.

For *lock-in*, the input signal V_S may be a 10^{-4} fraction taken from a cascaded voltage divider (Figure 16.3) biased by the reference sinusoidal signal V_R; basic tests are the following:
1) Linearity: proportionality between d.c. output and a.c. input amplitude.
2) Asynchronous signals rejection (by adding to V_S a signal taken from another oscillator or a d.c. voltage)
3) Phase response (using a phase shifter placed between V_R and the input voltage divider)
4) Evaluation of the useful frequency range, input amplitude range, quality factor.

To drive two analog switches in phase opposition, the circuit of Figure 16.9 may be used, where the potentiometer R trims the threshold voltage, in order to obtain symmetric square wave outputs (V^+, V^-) even if the reference voltage is affected by an offset voltage. The pull-up resistor (1 kΩ) is required only by open-collector comparators (as LM311/LF311 /μA710 /μA339).

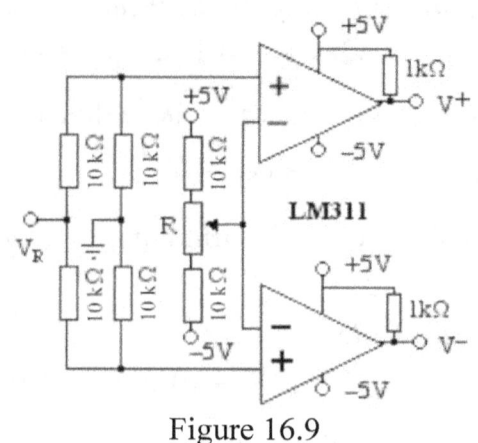

Figure 16.9

Figures 11.8, 11.9 and 11.10   Lock-in with multiplier ±1
OA= TL084, Analog switch: CD4016 or equivalent. $R_1 = R_o = R = 10$ kΩ, $C = 1$ μF, $f \approx 1$ kHz. Observe the signal before and after the low-pass filter (try with different time constants, e.g. change C). Insert a phase shifter (Figure 10.5) into the V_R channel and measure V_o as a function of the phase shift. Test the behaviour with different frequencies

Figure 11.12   Synchronous filter
$R = R_o = 100$ kΩ, $C = C_o = 100$ nF. Use a sinusoidal signal (some kHz) fed to V_R and to V_S ; add to V_S another (higher and lower frequency) noise and verify the rejection of asynchronous signal at the output. Compare signals before and after the high-pass filter. The output amplifier may have high gain ($G \approx 10 \div 50$).

In the following exercises on *logic circuits*, the default power supply is: +5 V /ground, valid for both TTL and CMOS. You may use four 1.5 V batteries in series to a diode ($V \approx 5.4$ V).
To check the logic value of any point of the circuit, use a LED connected to +5V through a 1 kΩ resistor (or 330 Ω for more bright signal): the value is "low" when LED is ON; this reverse logic is justified by the small output current of TTL in "high" state (about 0.4 mA, versus minimum LED current of about 2 mA). This inconvenience may be avoided using CMOS type 74HCxx or 74HC40xx.

Figure 12.2 , 12.3
Verify the De Morgam theorems using the truth tables (equivalence between NAND and OR with negated inputs, between NOR and AND with negated inputs).

Figure 12.5
Setup EXCLUSIVE OR using the three equivalent circuits and verify the truth table.

Figure 12.6 e 12.7
Build NOT, AND, OR, NOR, EX-OR first with NAND gates, then with NOR gates.

Figure 12.8

Build 6-input NOR with 7405 (Hex-inverter, open-collector) and pull-up 2.2 kΩ resistor connected to the 6 outputs.

Figure 12.9 e 12.10   RS Flip-Flop

Use 470 Ω resistors, (and 100 nF capacitances in Figure 12.10). Inverters: 7404, NAND:7400, NOR:7402. With reference to the truth tables, verify the *stable state* [row 4 in a) and row 1 in b)], starting from rows 2 or 3 (the outputs do not toggle). Verify the *disallowed state* (outputs both "high" for row 1 in a) and both "low" for row 4 in b).

Figure 12.11   Synchronous Flip-Flop

As CLOCK signal use the output of a RS flip-flop (Figure 12.10), to avoid switch bounces[90]. Verify that this circuit, with CLOCK enabled, is equivalent to the RS Flip-Flop of Figure 12.10a. Verify that this device does not toggle when Clock is "low". Repeat the exercise using four NOR gates.

Figure 12.12   Master-Slave flip-flop

Use first 9 NAND gates (7400), then 9 NOR gates (7402). Write the truth table for the second circuit, and ascertain whether the data are transferred during *rising* or *falling* edge in the CLOCK input. Use 6 LED to test the state of R,S, and of the outputs of master and slave.

Figure 12.13 Type-D flip-flop

Use two 7400 (NAND) and one 7404 (inverters). Try also the equivalent made wit NOR gates (7402). You may also use one of the two type-D flip-flop of 4013 or 7474 [91]

Figure 12.14   Divider-by-two

Use 7474. Try several cascaded stages and test the signals at each output

Figure 12.15   J-K flip-flop

Use two 7400 for the master-slave flip-flop and another 7400 for the two AND gates: two NAND (half 7400) with shorted inputs (inverters) in series to the other two NAND (NANDs becomes ANDs). Test the behavior with both J and K "high", then with both "low". Verify the equivalence to a type-D flip-flop when J is connected to K through an inverter. Try using 7473 as J-K flip-flop.

---

[90] See http://www.elexp.com/t_bounc.htm or http://www.labbookpages.co.uk/electronics/debounce.html
[91] Datasheet: http://www.doctronics.co.uk/4013.htm - about and http://www.ti.com/lit/ds/symlink/sn74ls74a.pdf

Figure 12.16   Type-T flip-flop
Use 7473 or 4027. Drive the CLOCK with a low frequency multivibrator and the TOGGLE input through a manual switch (between "low" and "high" levels). Obsserve the output at the oscilloscope.

Figure 12.17 and 12.18   Monostables
Use CMOS gates (4011, 4001): R = 10 kΩ pot in series to 1 kΩ, $R_1 = 100$ kΩ, $C = 10 \div 100$ nF, $C_1 = 1$ μF. Using TTL gates (7400, 7402), resistors R and R_1 should be smaller than 470 Ω to avoid an output latch to "high", due to the current fed to ground.

Figure 12.19
Use 7403, 4011 NAND gates, R = 100 kΩ, $C = 10 \div 100$ nF, a manual switch to drive input between "low" and "high" levels. Check the circuit behavior when replacing NAND with NOR gates (7402, 4001).

Figure 12.20, 22 and 23   Astables
Using TTL gates (7414 for 12.20 and 7402, 7404, 7408 for 12.22 e 12.23): R = 100 Ω pot in series to 100 Ω, C = 1 μF. Using CMOS gates (4584): R = 100 kΩ pot, C = 10 nF. Observe the output signal and the signal at the input of gate 1 while varying R.

Figure 12.24 and 12.25
Use 11 inverters, (or better 23 inverters) using four 74L04, or 7404, or 4069. Measure the frequency and the pulse width. Calculate the risetime of different gates as a function of bias voltage (2 V < V_{cc} < 5 V), and of the room temperature (use a soldering iron to heat the gates).

Figure 12.26   Delay generator
Use CMOS gates, and a square wave at the input (e.g. circuits of Figures 12.20 or 12.22), and choose a time constant RC smaller that the half-period of the square wave. Test several types of gate NAND, AND, OR, NOR (4011,4001,4071,4081).

Figure 13.3   A 555 monostable
Bias: +15 V, 0 V; $R_1 = R_2 = 10$ kΩ, $C_v = C_i = 10$ nF, C = 100 nF, R = 10 kΩ in series to a 1 MΩ pot. Drive the input with a pulser as that shown in Figure 9.6. Compare signals V_i and V_T, than V_i and V_c, then V_i and V_{out}, while varying R.

Figure 13.4   Astable pulser
$C_v = C = 10$ nF, $R_2 = 100$ kΩ, $R_1 = 1$ kΩ in series to a 100 kΩ pot. Observe the output signal while changing R_1. Replace R_2 with the circuit shown here, with

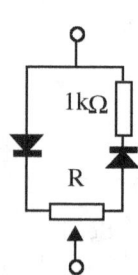

R =100 kΩ pot, and observe how the pot-setting does change the duty cycle (ratio T_2/T_1) without affecting the frequency. Test the pulser with different bias: e.g. +15 V/ 0 V and ±5 V.

Figure 13.5 Square wave generator
$R_1 = 1$ kΩ, $R_2 = 10$ kΩ in series to a 100 kΩ pot $C_v = C = 10$ nF, $R_L = 10$ kΩ pot. Observe the output signals 1 and 2, while changing the R_2 value. Displace the load R_L to the output 1 and see the change.

Figure 13.6 Linear voltage-to-frequency converter
OA = TL081, Timer = ICM7555. R = 100 kΩ, $R_o = 10$ kΩ, $C = C_o = 100$ nF. (R* = 1 kΩ to protect the trigger input). Observe the signals V_T and V_o while changing the voltage input. Try to change the values of τ and τ_o.

Appendix A

A simplified explanation of the transistor working principle in linear region

A complete analysis of the transistor reqires using complex models. Here we offer a very short description that allows understanding most part of common circuits involving transistors.

We only define, without explanation, the I-V characteristics for diode and transistor, and we use a reduced set of parameters (h_{fe} and h_{ie}) restricting our analysis to small a.c. signals and low frequencies (for d.c. signals one must take into account the junction bias voltages).

A.1. The diode

A diode is a two-terminal non-linear device made of a junction between two semiconductors, one P-doped and one N-doped. Near the junction the majority[92] carriers (holes in P-doped and electrons in N-doped) recombine, leaving a "*depletion*" layer [93] where there are no free charge carriers. The charges bound to the crystal lattice (positive ions in N-doped material and negative ions in P-doped material), are fixed, so the electric field, due to this double layer of opposite charge, limits the diffusion of the majority carriers (electrons frm N region and holes from P region). This electric fiels, however, does not block the flux of minority carriers thermally generated inside the depletion layer (eletcrons flowing from P to N and holes from N to P). This *inverse* current is nusually named I_o.

In equilibrium conditions (without external voltage applied to the diode) the inverse current is balanced by the forward current I_d, due to the diffusion of majority carriers: $I = I_d + I_o = 0$.

The forward current I_d depends on the applied voltage V, on the absolute temperature T and on the type of semiconductor as follows:

$$I_d = C \exp[qV/(nK_BT)] \qquad (A.1)$$

where q is the elementary charge, K_B the Boltzman constant, and *n* a constant that is about 1 for germanium and about 2 for silicon. Letting $K_BT/q = V_t$ (at room temperature $V_t \approx 26\,mV$) we get $I(V) = C\, e^{V/nV_t} - I_o$.

The constant C is determined by the equilibrium condition: $I(0) = 0$ that gives $C = I_o$, so that we may write (A.1) as the *ideal diode equation:* :

$$I(V) = I_o (e^{V/nV_t} - 1). \qquad (A.2)$$

---

[92] For basic definitions of doped semiconductors, holes, majoriry and minority carriers, depletion layer,.... see for example: *Elementary Semiconductor Physics*, H.C. Wrigth, or *The Physics of Semiconductor Devices*, D.A. Fraser, or *Introduction to Semiconductor Physics*, R. Adler, A. Smith, R. Longini, or *Semiconductor Devices*, S.M. Sze.

[93] The depletion layer thickness is of the order of μm.; see http://en.wikipedia.org/wiki/P–n_junction

For $V \gg V_t$ the constant 1 in (A.2) may be neglected with respect to the exponential term:

$$I(V) \approx I_d = I_o \, e^{V/nV_t}, \quad V \gg V_t. \qquad (A.3)$$

The slope of the characteristic curve (Figure A 1a) is $\partial I/\partial V = I/(nV_t) = (r_d)^{-1}$, ove r_d is named *dynamic resistance* of the forward biased diode, that increases linearly with T: at room temperature $r_d \approx 26/I \, (\Omega/mA)$.

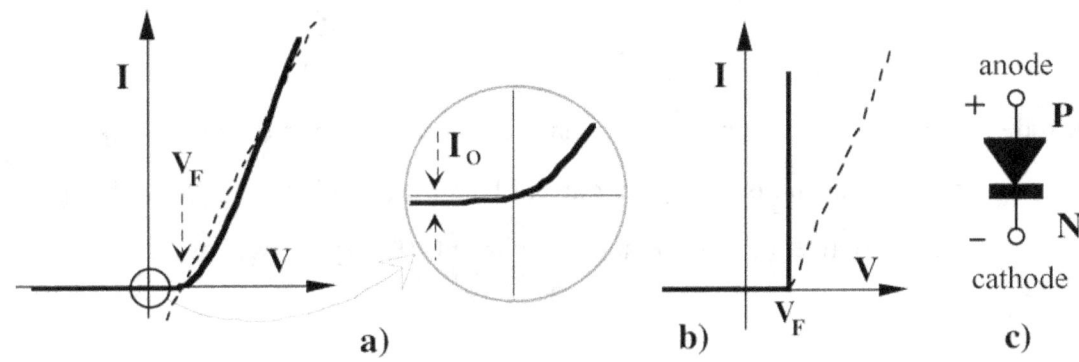

Figure A 1

For $V < 0$ the exponential term becomes negligible, so that $I(V<0) \approx -I_o$.

The reverse current I_o depends on the specific diode, and it is normally quite small, of the order of 1 µA. We may conclude that the diode is essentially a rectifier: it may in fact be approximated as *unidirectional switch*. In Figure A1b the characteristic curve is approximated by a piecewise linear function (the dotted line defined by: $I \approx 0$ for $V < V_F$, and $I \approx (V - V_F)/r_d$, for $V > V_F$), where for germanium diodes $V_F \approx 0.6 \, V$ and for silicon diodes $V_F \approx 0.2 \, V$.

More often, beside neglecting the reverse current, also the dynamical resistance is neglected, which leads to the *ideal unidirectional switch* model: when forward biased the diode is assumed to be a voltage source V_F, when reverse biased it is assumed to be an open switch (Figure A1b). This model is illustrated in Figure A 2 where the series of the diode and the resistor R (with $R \gg r_d$) is a rectifying voltage divider: the input a.c. signal V_i appears at the output without the negative half-wave (V_u).

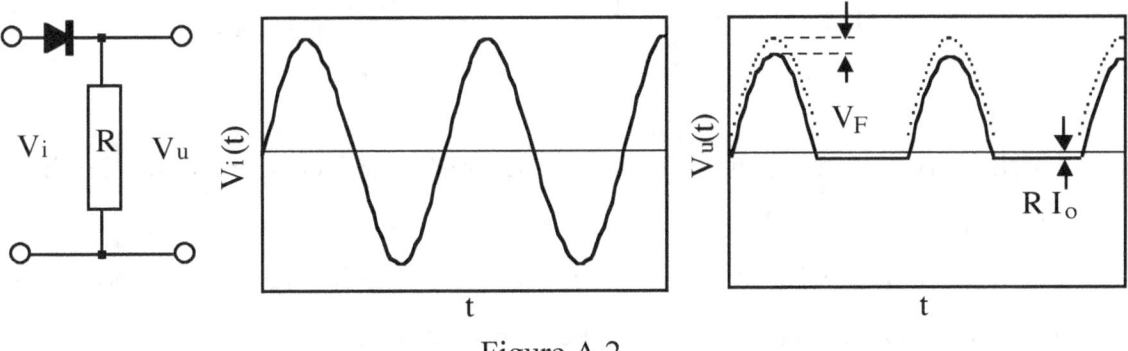

Figure A 2

In the circuit of Figure A.3 the capacitor placed in parallel to the resistor, is charged during the positive half-wave through r_d and discharged during the negative half-wave through R.

If $R \gg r_d$, the output signal is that shown by the full line (here the effect of V_F is neglected).

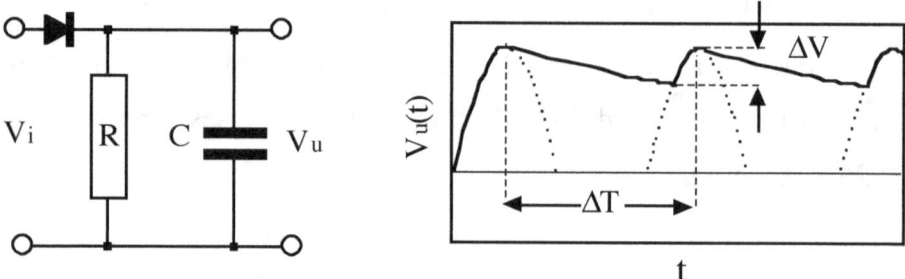

Figure A 3

The amplitude ΔV of the ouput signal in stationary conditions, is name *ripple*, may be calculated as follows. Assuming a constant discharging current $I = V_u/R$, if $\Delta T = 1/f$ is the signal period from the definition of capacitance ($C = q/V$) and of current ($I = \partial q/\partial t$) we get $\Delta V = V\Delta T/RC$, or $\Delta V/V = (fRC)^{-1}$.

A.2. The zener diode

The current flowing across a reverse biased diode is normally very small, even considering the small leakage current due to the surface conductivity (increasing $|-V|$) that is added to the reverse current I_o. However, when the reverse voltage reaches the *breakdown voltage* V_B, whose value depends on the particular diode, a different process occurs: the *avalanche* conduction. The high electric field, within the depletion layer, gives to the electrons enough energy to generate, by collision, new charge-carrier pairs. This phenomenon leads eventually to the junction distruction, when the local power dissipation exceeds a limit value.

Some diodes, named *zener* diodes, are specially manufactured to withstand high reverse voltages without damage.

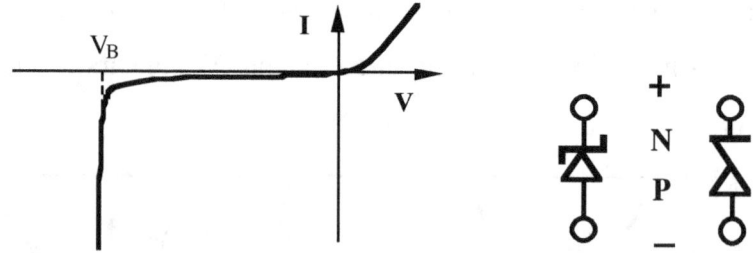

Figure A 4

The characteristic curve of a zener diode, and the zener graphic symbols, are shown in Figure A4, where $+-$ signs mark the *reverse* bias.

The zener diode may be used as voltage stabilizer: for example Figure A 5 shows how the amplitude of the input signal ΔV_i is reduced in the output ripple ΔV_z.

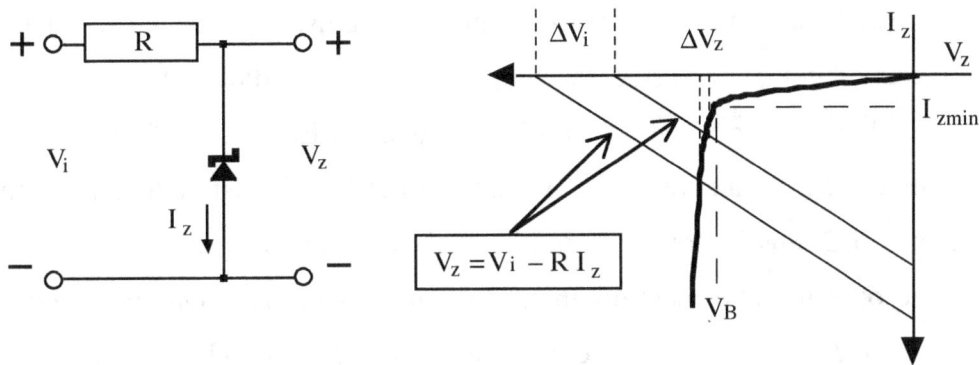

Figure A 5

In Figure A5 the axes V_z and I_z directions are reverted, so that the reverse voltage and the current are positive. The *load line* is $V_z = V_i - RI_z$. The slope of the load line does slightly change with the load resistance R_L placed in parallel to the zener, (i.e with the total current $I = I_z + V_z/R_L$ flowing across R), but the change in V_z remains small, until $V_i - RI > V_B$. The maximum current I_L that can be fed to the load is $I_L = V_z/R_L < (V_i - V_z)/R + I_{Zmin}$, where I_{Zmin} is the zener current at the voltage V_B.

A.3. The transistor : some definitions

The transistor[94] is a three-terminal device (collector, base and emitter) made of two p-n junctions in series, as shown in Figure A 6.

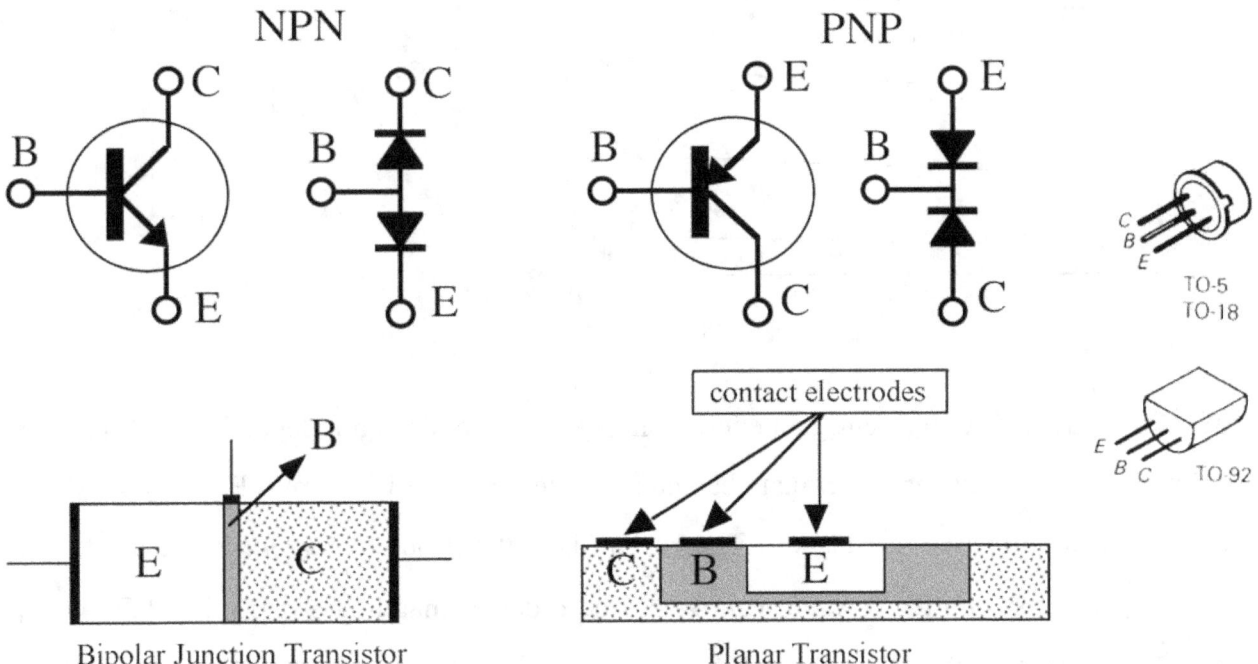

Figure A 6

When the two anodes are common we have a NPN transistor, when the two cathodes rae common

---

[94] See http://en.wikipedia.org/wiki/Transistor and http://en.wikipedia.org/wiki/Bipolar_junction_transistor

we have a PNP transistor. The two junctions, however must be very close each other and interacting (we cannot get a transistor by simply joining two diodes!). The charge carriers injected into the depletion layer of the forward biased junction EB must diffuse into the depletion layer of the reverse biased junction BC: in other words the diffusion length of the charges injected into junction EB must be longer than the junction thickness[95].

We'll here analyze only the two most common configurations: the *common-emitter* (amplifier) and the *common-collector* (follower), in a.c. regime. We'll study the NPN transistor; for the PNP the analysis is similar, with reverses bias.

A.4. Common emitter

The transistor *linear region*[96], also named *active region*, is a limited area in the I_c, V_{ce} plane, as shown in Figure A7, that gives an example of the characteristic curves $I_c(V_{ce}, I_b)$ of the collector current I_c versus the collector-emitter voltage V_{ce}, for several values of the base current I_b.

In the linear region I_c has a weak dependence on V_{ce}, so that, for each I_b value, the $I_c = I_c(V_{ce})$ curves may be approximated by horizontal segments.

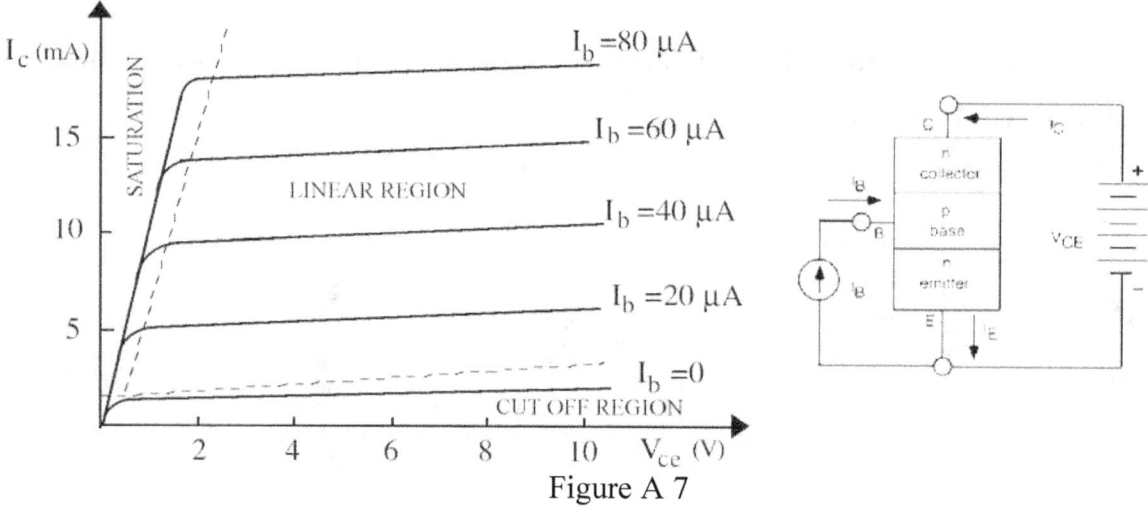

Figure A 7

Therefore we may define a current-gain coefficient $\beta = I_c/I_b$, that does not depend on I_b (in a first approximation). A second parameter that characterizes the transistor is the ratio $R_b = v_{be}/i_b$, which is the BE-junction effective resistance[97]. The current i_b is the *dynamic current* injected into the base and v_{be} the base-emitter *dynamic voltage*[98]. The order of magnitude of R_b is 1 kΩ, and β varies for different transistors from 20 to 300.

---

[95] When both BE and BC are forward biased the transistor is in the *saturation region*, when both are reverse biased the transistor is in the *cut-off region*.
[96] The transistor *linear region* must not be confused with the OA linear region.
[97] In the four-parameter model of common-emitter configuration: $\beta = h_{fe}$ and $R_b = h_{ie}$.
[98] Dynamic current and dynamic voltage are defined in §A.5

The fact that $\beta \gg 1$ may be explained by the following arguments (for NPN transistor). The BE junction is forward biased and therefore the majority charge carriers in the emitter (electrons) flow from E to B: most part of these electrons diffuse into the depletion layer of the BC junction that is reverse biased. This charge flux, modulated by the BE bias voltage, adds to the BC reverse current. Here also an avalanche current multiplication may occur, due to the high reverse bias, leading to an increase current gain.

A more detailed treatment of this complex phenomenon may be found elsewere [99].

A.5 Dynamic regime

Let us assume that the transistor in Figure A8a is biased within the linear region.

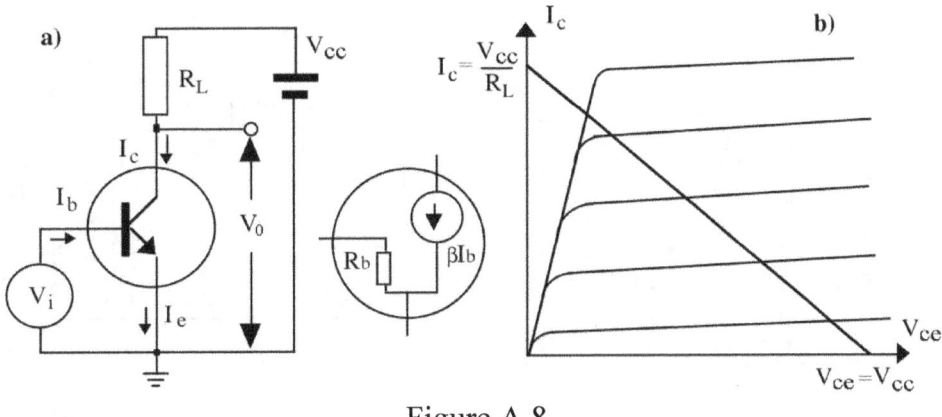

Figure A 8

In Figure A8.b the load line is defined as $V_{ce} = V_{cc} - R_L I_c$. To each value of the input voltage V_i corresponds a different value for the base current I_b, i.e a different characteristic curve: the collector current $I_c (V_i)$ is determined by the crossing between each curve with the load line.

When the input voltage changes, the working point moves along the load line, thus changing the output voltage $V_o = V_{ce}$.

We define as *dynamic voltages* and *dynamic currenst* the *changes* of voltage and current, respectively, with respect to the values taken for a given position of the working point on the load line (*quiescent* point, or *Q-point*[100]). These variables will be written here in low-case $v_i = V_i - V_{iQ}$, $v_o = V_o - V_{oQ}$, $i_b = I_b - I_{bQ}$, $i_c = I_c - I_{cQ}$, etc. In this way we may neglect in our analysis the contribution of constant terms (as bias voltages or juntion voltage drops).

The output dynamic voltage v_o is obtained by differentiating the load line equation:

$$v_o = \Delta V_{ce} = \Delta(V_{cc} - R_L I_c) = -R_L \Delta I_c = -R_L i_c.$$

---

[99] See http://en.wikipedia.org/wiki/Bipolar_junction_transistor and references therein
[100] See http://en.wikipedia.org/wiki/Q-point or http://en.wikipedia.org/wiki/Biasing

From the definition of voltage gain $A_V = v_o/v_i$, and from $v_i = R_b i_b$, we get:

$$A_V = -(R_L i_c)/(R_b i_b) = -\beta R_L/R_b,$$

The voltage gain A_V depends on the transistor parameters β and R_b (which, in turn, depend on the temperature T). However it is possible to remove the gain dependance on β and R_b simply adding a resistor R_E in series to the emitter, as in Figure A9.

In Figure A9a the voltage divider (R_1, R_2) sets the Q-point of the transistor, and the input voltage is applied through a capacitor, to avoid the effects of the input source on the transistor bias. If we assume this capacitor to be large enough, we may neglect its impedance[101]. Figure A9b shows the equivalent dynamic circuit and explicits the BE-junction effective resistance and the current controlled current source βi_b

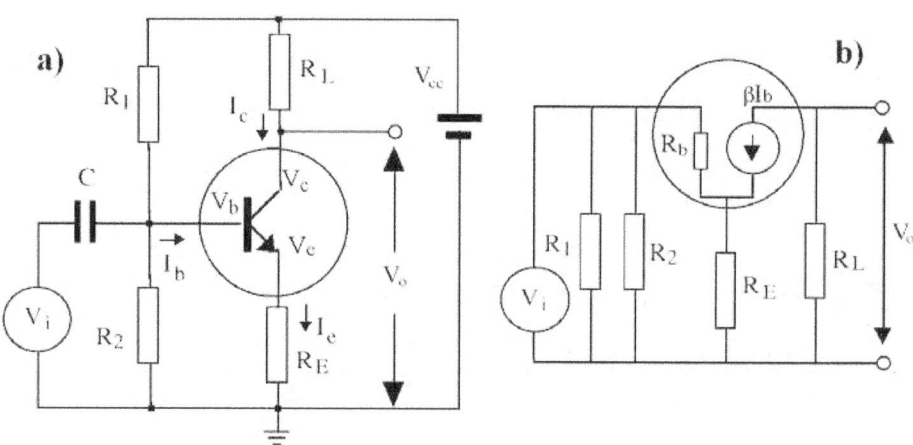

Figure A 9

The input dynamic voltage v_i may be written:

$v_i = R_b i_b + R_e i_E = R_b i_b + R_E (1+\beta) i_b = [R_b + R_E (1+\beta)] i_b,$

So that the input impedance (neglecting the bias voltage divider and the capacitor) is:

$$Z_i = v_i/i_b = R_b + R_E (1+\beta).$$

Because $\beta \gg 1$ and $R_b \approx 1\,k\Omega$, letting $R_E \approx 1\,k\Omega$ we may neglect R_b with respect to $(1+\beta)R_E$.

Therefore: $Z_i \approx (1+\beta)R_E \approx \beta R_E$, i.e. the input impedance is approximately β times the emitter resistance R_E.

The output dynamic voltage is $v_o = -R_L i_c = -\beta R_L i_b$, and therefore the voltage gain is:

$$A_V = \frac{V_u}{V_i} = -\frac{\beta R_L i_b}{R_i i_b} = -\beta \frac{R_L}{R_i} \approx -\frac{\beta}{1+\beta} \frac{R_L}{R_E} \approx -\frac{R_L}{R_E}$$

which does not depend on the particular transistor used.

Taking into account the voltage divider (R_1, R_2) the effective input impedance becomes:

$$Z_i = R_1 \| R_2 \| \beta R_E = (1/R_1 + 1/R_2 + 1/\beta R_E)^{-1},$$

---

[101] If we work at frequency ω, must be $C \gg 1/\omega(R_1 \| R_2)$ (see § 5 and Appendix B).

and, if the smaller resistance between R_1 and R_2 is of the order of R_E, we get $Z_i \approx R_1 \| R_2$.

The output impedance Z_o is defined by the ratio $v_o\{i_o=0\}/i_o\{v_o=0\}$, where $v_o\{i_o=0\} = R_L i_c$, and $i_o\{v_o=0\} = i_c = \beta i_b$. In conclusion: $Z_o = R_L i_c / i_c = R_L$.

In order to optimize the resistance values in the voltage divider R_1, R_2 we note that:

1) to maximize the output voltage range we should choose the quiescent point at $V_{oQ} \approx V_{cc}/2$; this defines the collector current I_{cQ};

2) the given I_{cQ} defines the emitter voltage $V_{eQ} = R_E I_{cQ}$. To keep the transistor inside the *active region* must be $V_{bQ} > V_{eQ}$, i.e. $V_{bQ} > V_{eQ} + V_{be} \approx V_{eQ} + 0.6V$;

3) the values R_1, R_2 cannot be too high because we need to keep the base current negligible with respect to the current flowing across the divider: $V_{cc}/(R_1+R_2) \gg I_b$.

A.6. Common collector (Emitter Follower)

In the common-collector configuration (Figure A 10), the output is taken at the transistor emitter, with the collector connected at the common voltage V_{cc}.

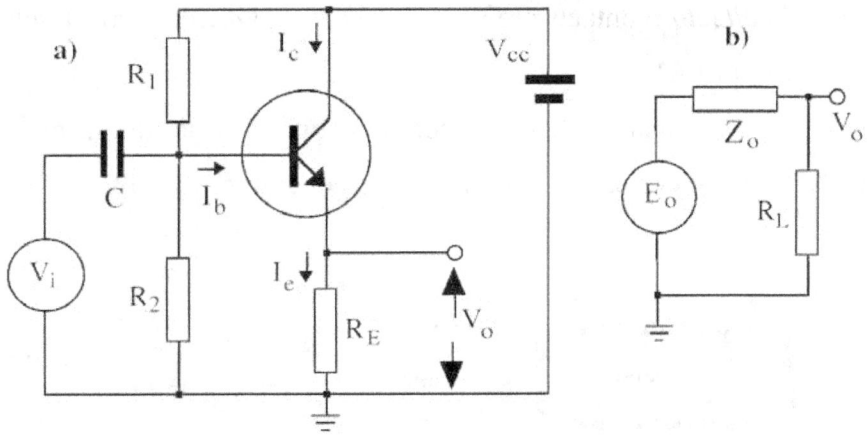

Figure A 10

Here the load line equation is $V_{ce} = V_{cc} - R_E i_e$, and the dynamic ouput voltage (*not loaded*) is :

$$v_o = R_E i_e = R_E (1+\beta)] i_b.$$

The input dynamic voltage is $v_i = R_b i_b + R_E i_e = [R_b + R_E(1+\beta)] i_b$, so that the voltage gain is:

$$A_V = \frac{v_o}{v_i} = 1 / \left(1 + \frac{R_b}{(1+\beta)R_E}\right).$$

For $(1+\beta)R_E \gg R_b$, we get[102] $A_V \approx 1$, showing that the output voltage follows the input voltage [103].

The input impedance Z_i is:

---

[102] For too small values $R_E < R_b/(1+\beta)$, also the gain decreases: $A_V \approx [(1+\beta)R_E/R_b] < 1$.
[103] Note that the d.c. level of the emitter voltage is $V_b - V_{be} \approx V_b - 0.6$ V.

$$Z_i \approx (1+\beta)R_E \approx \beta R_E,$$

(or, accounting for the bias voltage divider) $Z_i \approx R_1 \| R_2 \| \beta R_E$.

Figure A10b shows that the output impedance Z_o is the value that, applied as a load for the ideal voltage generator $E_o = v_o\{R_L = \infty\} = v_i A_V$, halves the output voltage: i.e. $v_o\{R_L = Z_o\} = E_o/2$. This last equation may be written as:

$$v_o\{R_L = Z_o\} = \frac{v_i}{1 + R_b/[(1+\beta)(Z_o \| R_E)]} = E_o/2 = \frac{v_i}{1 + R_b/[(1+\beta)R_E]}$$

With some algebra we get $Z_o = R_E \| [R_b/(1+\beta)] \ll R_E$.

As a conclusion: $Z_o/Z_i \approx R_b/R_E \, \beta^2 \ll 1$. The common collector behaves as a *current amplifier*.

A.7. Field Effect Transistor (FET)

The transistor described in §A.3– §A.6 is the *Bipolar Junction Transistor* (BJT), where the collector and emitter currents are controlled by the base current (*current-controlled* device). Other transistors are instead *voltage-controlled* devices, e.g. the Field Effect Transistors (FET), where the *emitter* and *collector* contacts are named *source* and *drain*, respectively, and the *base* is replaced by the *gate* (Figure A11).

The current flowing in the channel, that connects the source to the drain, cis controlled by the gate voltage, and the leakage current trough the gate is generally negligible (of the order of nA).

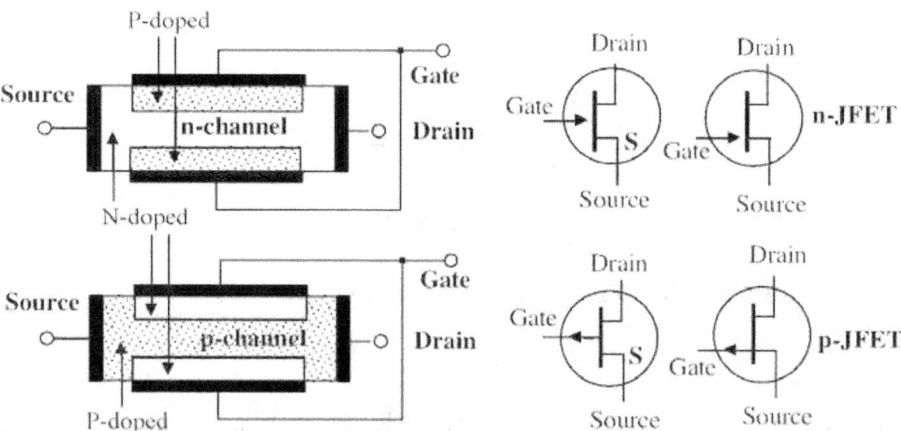

Figure A 11

FET transistors may be classified in two classes: the Junction Field Effect Transistor (JFET) where the control voltage is applied through a reverse biased junction, and the Metal-Oxide-Semiconductor FET (MOSFET) where the control voltage is applied through an insulating layer. Within each class we distinguish between n-type channel (n-JFET, n-MOSFET), and p-type channel (p-JFET, p-MOSFET). The integrated circuits made of MOSFET are named *Complementary*-MOSFET (CMOS).

Drain and source terminals are almost equivalent, however the source is normally marked as S or it is drawn closer to gate, in the graphic symbols.

Figure A 12 shows the MOSFET structure and symbols. The working principle for both types is based on the change of channel cross-section induced by the voltage applied to the gate with respect to the substrate.

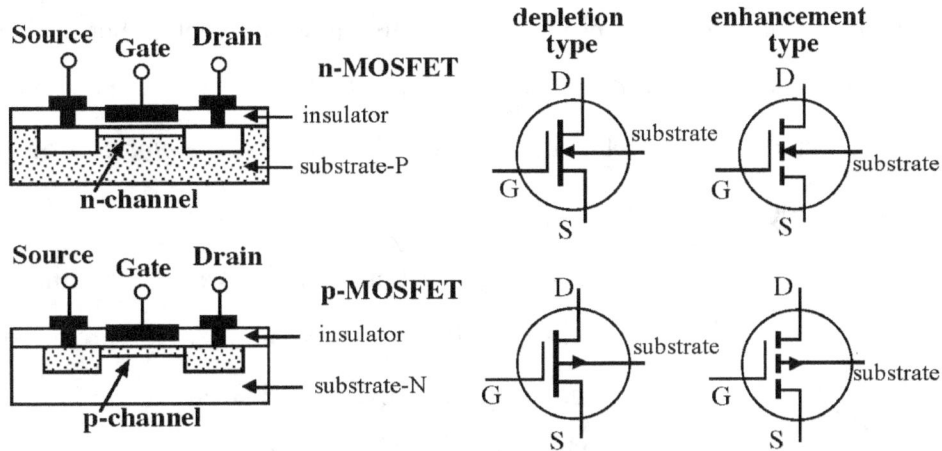

Figure A 12

In JFET the voltage applied to the gate must never forward bias the gate-channel junction, while in MOSFET the applied voltage is limited only to values that do not produce damage to the thin insulation layer: for example electrostatic charge build-up may destroy the device.

In JFET with grounded source the gate voltage increases the channel current when it approaces the drain voltage gate: $\partial I/\partial V_{GS}$ is positive for n-JFET and negative for p-JFET.

Therefore an n-JFET resembles a npn BJT, and p-JFET resembles a pnp BJT. The input impedance of JFET is much higher than for BJT, of the order of $10^{10} \div 10^{14}$ Ω.

MOSFET may be of different types: *depletion mode* (channel conducting with $V_{GS}=0$), or *enhancement mode* (channel off with $V_{GS}=0$). In the first case the channel is doped with the same sign as drain and source, but with weaker doping; in the secon case the channel is generated by the bias that produces an *inversion layer* close to the gate, and the conduction begins at a threshold value $V_{GS}=V_T$[104].

---

[104] For details http://en.wikipedia.org/wiki/MOSFET or *Microelectronics*, J. Millman et al.

Appendix B

B.1. Complex numbers

A complex number[105] can be viewed as a point or position vector in a two-dimensional Cartesian coordinate system called the *complex plane*. We name *imaginary* the axis y, and we name *real* the axis x. We may associate to every point (a, b) of the plane the vector that projects the origin (0,0) into (a,b). This vector represents the *complex number* $C = a + jb$. The symbol j is the imaginary unit ($j = \sqrt{-1}$ and $j^2 = -1$). Let m be the vector length: it is the *modulus* of the associated complex number C: $m = |C| = \sqrt{a^2 + b^2}$.

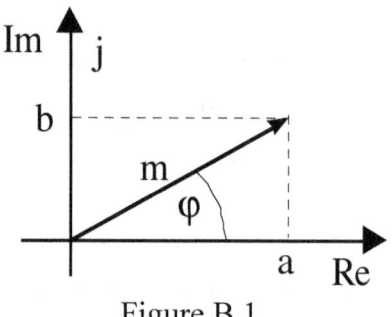

Figure B 1

The angle φ between the vector and the x axis is the *phase* of the complex number: $\varphi = \arctan(b/a)$, and we may write $C = m(\cos\varphi + j\sin\varphi)$.

The *real part* of C is $a = m\cos\varphi = \text{Re}(C)$, and the *imaginary part* of C is $b = m\sin\varphi = \text{Im}(C)$.

From the Euler formula[106]:

$$\exp(j\varphi) = \cos\varphi + j\sin\varphi,$$

we get $C = m e^{j\varphi}$. The *complex conjugate* of C is the number $\overline{C}$ (or C*) with the same real part and imaginary part with opposite sign: $C^* = a - jb = m e^{-j\varphi}$.

The *sum* of two complex numbers is:

$$C_3 = C_1 + C_2 = (a_1 + jb_1) + (a_2 + jb_2) = (a_1 + a_2) + j(b_1 + b_2) = a_3 + jb_3,$$

with $a_3 = a_1 + a_2$ and $b_3 = b_1 + b_2$.

The *product* of two complex numbers is:

$$C_3 = C_1 \times C_2 = (m_1 e^{j\varphi_1}) \times (m_2 e^{j\varphi_2}) = m_1 m_2 e^{j(\varphi_1 + \varphi_2)} = m_3 e^{j\varphi_3},$$

with $m_3 = m_1 m_2$ and $\varphi_3 = \varphi_1 + \varphi_2$.

The quotient of two complex numbers is:

$$C_3 = C_1 / C_2 = (m_1 / m_2) e^{j(\varphi_1 - \varphi_2)}.$$

B.2. Sinusoidal voltages and currents in complex notation

Let us consider a sinusoidal current signal: $i(t) = I_o \cos\omega t$, where $\omega = 2\pi\nu$ is the angular frequency

---

[105] See http://en.wikipedia.org/wiki/Complex_number
[106] See http://en.wikipedia.org/wiki/Euler%27s_identity

and ν is the frequency. We associate i(t) to the complex number $I=I_0(\cos\omega t+j\sin\omega t)$, or, in the *polar form* $I=I_0\exp(j\omega t)$.

Frequently the notation is simplified by using the complex variable $s=\sigma+j\omega$ (with σ and ω real), and letting $I(s)=I_0 e^{st}$. For voltages we similarly let $V(s)=V_0 e^{st}$. Note that this notation, assumed for sinusoidal signals, is fully general, because any signal may be represented by a (finite or infinite) series of sinusoidal componebts (Fourier representation[107]).

B.3. Complex impedance[108]

Resistors, capacitors and inductors are linear elements that associate the current i(t) to the voltage drop v(t) across them (where t is the time variable).

The relation $v(t)=\{Z\}i(t)$ defines the impedance {Z} *operator* that transforms the current i(t) into the voltage v(t). For a resistor the impedance is a real constant

$$v(t) = R\,i(t) \qquad \rightarrow \qquad \{Z_R\} = \{R\}.$$

For a capacitor, from the capacitance definition $C=q/v$ and the current definition $i=\partial q/\partial t$, we get $i(t)=C\,\partial q(t)/\partial t$, or by integration:

$$v(t) = (1/C)\int i(t)\,dt \qquad \rightarrow \qquad \{Z_C\} = \{(1/C)\int dt\}.$$

For an inductor the voltage induced by the magnetic flux variation due to current changes is:

$$v(t) = L\,\partial i(t)/\partial t \qquad \rightarrow \qquad \{Z_L\} = \{L\partial/\partial t\}.$$

In the simplified complex notation the corresponding transformations are:

$$\{Z_R\}I(s) = R\,I(s) \qquad \rightarrow \qquad \{Z_R\} = R$$

$$\{Z_C\}I(s) = (I_0/C)[\int e^{st}\,dt] = (1/sC)\,I(s) \qquad \rightarrow \qquad \{Z_C\} = 1/sC = 1/j\omega C$$

$$\{Z_L\}I(s) = (I_0 L)[\partial e^{st}/\partial t] = (sL)\,I(s) \qquad \rightarrow \qquad \{Z_L\} = sL = j\omega L$$

remembering that $\int e^{st}dt = e^{st}/s$ and $\partial(e^{st})/\partial t = s\,e^{st}$.

Therefore the voltage drop across a resistor, a capacitor or an inductor may be written as:

$$V_R(s) = R\,I(s)$$
$$V_C(s) = (1/sC)\,I(s)$$
$$V_L(s) = (sL)\,I(s)$$

The same results are obtained by using the trigonometric notation: $I(\omega t)=I_0(\cos\omega t+j\sin\omega t)$, remembering $\int\cos(x)\,dx=\sin(x)$, $\int\sin(x)=-\cos(x)$, $\partial\cos(x)/\partial x=-\sin(x)$, $\partial\sin(x)/\partial x=\cos(x)$, and $j^2=-1$.

---

[107] See http://cnyack.homestead.com/files/afourse/fsdef.htm
[108] See http://en.wikipedia.org/wiki/Electrical_impedance

B.4. Complex transfer function

Any linear network may be seen as a quadrupole, defined by the complex transfer function $T(s) = A(\omega) e^{j\phi(\omega)}$ (with modulus A and phase ϕ), that transforms the input complex signal $V_i(s)$ into the output complex signal $V_u(s) = T(s) V_i(s)$.

In a linear network the transfer function may be always written as ratio between two polynomes: $T(s) = N(s)/D(s)$. The roots (z_1, z_2, ...) of the numerator $N(s)$, i.e. the solutions of the equation $N(s) = 0$, are named *zeros* (for $s = z_i$, $T = 0$), and the roots (p_1, p_2, ...) of $D(s)$ are named *poles* (for $s = p_i$, T diverges).

Therefore we may always write:

$$T(s) = G \frac{(s-z_1)(s-z_2)(s-z_3)...}{(s-p_1)(s-p_2)(s-p_3)...}$$

Poles of a transfer function and stability criteria.

If the *real part of all poles is negative*, then the overall system is *stable*. If one *pole has a zero real part*, then that component is *critically stable*. If one *pole has a positive real part*, then that component leads the overall system to *instability*. If the imaginary part of a pole is zero, then that component does not have any oscillatory contribution. If the *imaginary part is not zero* then *its value is the frequency of oscillation* of the corresponding component of the system. The zeros of a transfer function do not affect the stability, they affect the transient response of the system.

B.5. Bode diagrams of a transfer function

The Bode diagrams of a transfer function are *piecewise linear approximations* of the curves $A(\omega)$ and $\phi(\omega)$ in *bi-logarithmic* and *semi-logarithmic* plots, respectively.

Let us consider two simple examples in sinusoidal regime.

1) Low-pass L-C filter (Figure B2). We have $V_u(s) = R I(s)$ and $V_i(s) = (R + Z_L) I(s)$ so that the transfer function is: $T(s) = V_u(s)/V_i(s) = R/(R+Z_L) = 1/(1+j\omega L/R) = 1/(1+j\omega/\omega_0)$. Therefore $A(\omega) = (1 + \omega^2/\omega_0^2)^{-1/2}$, and $\phi(\omega) = -\arctan(\omega/\omega_0)$. The piecewise linear approximation may be performed in two frequency regions: for $\omega \ll \omega_0$, where $\omega_0 = R/L$ is the break frequency, we may approximate $A = 1$, and $\phi = 0$, at $\omega = \omega_0$ $A = -\log\sqrt{2} \approx 0$, and $\phi = -\pi/4$, while for $\omega \gg \omega_0$ we may approximate $A = \omega_0/\omega$, and $\phi = -\pi/2$.

Therefore the Bode plot of $A(\omega)$ is made by the straight line $y = \log|T| = 0$ and by the straight line $y = \log|T| = \log(R/L) - \log\omega$, while Bode plot of $\phi(\omega)$ is made by the two horizontal lines

$\phi=0$ and $\phi=-\pi/2$, joined by the segment passing through the point $(\omega_0,-\pi/4)$, with a slope -45° /decade, in fact $\phi(\omega_0/10)=-\arctan(0.1)\approx 0$, and $\phi(10\omega_0)=-\arctan(10)\approx -90°$.

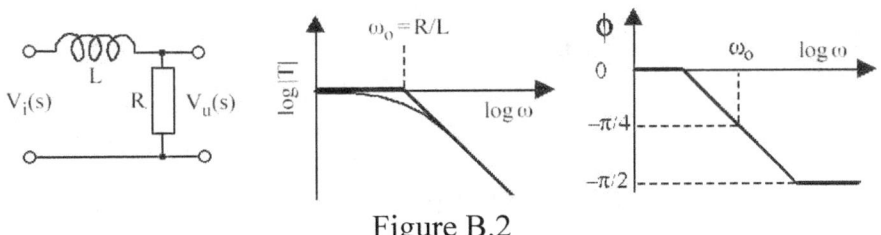

Figure B.2

2) High-pass RL filter (Figure B3). Here $V_u(s)=Z_L I(s)$ and $V_i(s)=(Z_L+R) I(s)$ so that the transfer function is $T(s) = sL/(R+sL) = 1/(1+j\,\omega_0/\omega)$. Therefore $A(\omega)=(1+\omega_0^2/\omega^2)^{-1/2}$, and $\phi(\omega)=\arctan(\omega_0/\omega)$. For $\omega \ll \omega_0 = R/L$ we may approximate $A=\omega/\omega_0$, and $\phi=+\pi/2$, at $\omega=\omega_0$ $A=-\log\sqrt{2}\approx 0$, and $\phi=+\pi/4$, while for $\omega \gg \omega_0$ may approximate $A=1$, and $\phi=0$. Therefore the bode diagram for $A(\omega)$ is made by the straight line $y=\log|T|=\log(R/L)+\log\omega$ at low frequencies and by the straight line $y=\log|T|=0$ at high frequencies. The phase Bode diagram is made by the two horizontal lines ($\phi=+\pi/2$ and $\phi=0$), joined by the segment passing through the point $(\omega_0,+\pi/4)$, with a slope of $-45°$/decade.

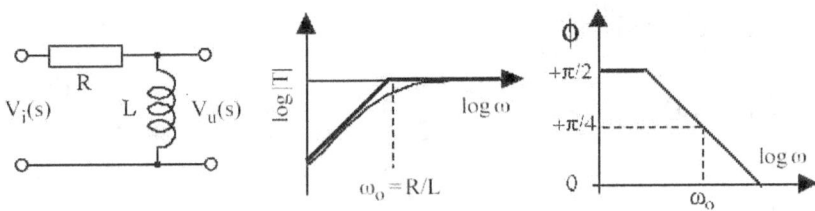

Figure B.3

B.6. Laplace Transform

Last century, before the advent of electronic calculators, people used the *slide rule* (also known as *slipstick*) to easier perform multiplications and divisions. The trick was to make a logarithmic *numerical transformation*, then to use sum and subtraction, then to make the inverse (anti-logarithmic) *numerical trasformation*. A similar technique may be applied to *functions*, instead of *numbers*, using Fourier or Laplace transformations[109].

The Laplace transform L, is applied to a real function f(t), using the following definition:

$$F(s)=L[f(t)] = \int_0^\infty [e^{-st}f(t)]dt,$$

where s is a complex variable and t is a real variable (time).

---

[109] See http://cnyack.homestead.com/files/idxpages.htm and http://en.wikipedia.org/wiki/Laplace_transform and http://www.stanford.edu/~boyd/ee102/laplace.pdf

It may be proven [110] that (if $f(t)=0$ for $t<0$) the following relations hold:

$$L[f_1(t)+f_2(t)] = L[f_1(t)]+L[f_1(t)] = F_1(s)+F_2(s)$$

$$L[af(t)] = aL[f(t)] = aF(s)$$

$$L[f(t-t_o)] = e^{-st_o}L[f(t)] = e^{-st_o}F(s)$$

$$L[\partial f(t)/\partial t] = sL[f(t)] - f(0) = sF(s) - f(0)$$

$$L\left[\int_0^t f(x)dx\right] = \frac{1}{s}L[f(t)] = F(s)/s$$

$$f(\infty) = \lim_{s \to 0}\{sL[f(t)]\} = \lim_{s \to 0}\{sF(s)\}$$

$$f(0) = \lim_{s \to \infty}\{sL[f(t)]\} = \lim_{s \to \infty}\{sF(s)\}$$

Moreover if $u(t)$ is the unitary *step function*, defined by: $u(t)=0$ for $t<0$ and $u(t)=1$ for $t>0$:

$$L[u(t)] = \int_0^\infty (1\ e^{-st})dt = 1/s$$

and for the exponential function:

$$L[\exp(at)] = \int_0^\infty e^{(a-s)t}dt = 1/(s-a).$$

From the Euler relation we get also:

$$L[\sin(at)] = \frac{1}{2j}\left(\frac{1}{s-ja} - \frac{1}{s+ja}\right) = a/(s^2+a^2)$$

$$L[\cos(at)] = \frac{1}{2}\left(\frac{1}{s-ja} + \frac{1}{s+ja}\right) = s/(s^2+a^2).$$

To show how to use the Laplace transform in electronics we analyze some simple examples. Let us first consider the response of an RC high-pass to an input step function.

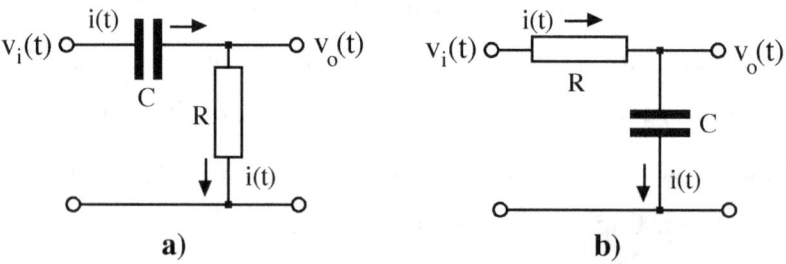

Figure B.4

With reference to Figure B4a and from the definition of current $i(t)=\partial q/\partial t$ and capacitance $C=q(t)/v(t)$, we obtain the equation:

---

[110] A short treatment may be found in *Electronics for the Physicist*, C.G. Delaney, chapt. 12.

$$v_i(t) = v_C(t) + v_R(t) = (1/C)\int i(t)dt + Ri(t). \qquad [b1]$$

With the Laplace transformation we get the complex equation, for generic functions v(t) and i(t):

$$V_i(s) = (1/sC)\,I(s) + RI(s) = (R + 1/sC)I(s). \qquad [b2]$$

On the other hand the output voltage may be written:

$$v_o(t) = R\,i(t) = R\,\{L^{-1}[I(s)]\} = L^{-1}\left[V_i(s)\frac{sRC}{1+sRC}\right]. \qquad [b3]$$

where L^{-1} is the Laplace inverse-transform. The output voltage may be written also:

$$v_o(t) = L^{-1}\,[L\{v_i(t)\}\,T(s)]. \qquad [b4]$$

For an input step function with amplitude V, we get $V_i(s) = L\,[V \cdot u(t)] = V/s$, and remembering that $L^{-1}\,[1/(s+a)] = \exp(-at)$, relation [b3] becomes:

$$v_o(t) = L^{-1}[V/(s + 1/RC)] = V\,\exp(-t/RC). \qquad [b5]$$

The same result may be obtained from [b4] that is a general relation.

Using this shortcut we analyze the case of the low-pass filter of Figure B4b, whose trasfre function is $T(s) = (1/RC)/(s + 1/RC)$, again using the step function for the input signal $V_i(s) = V/s$. From [b4] we obtain immediately:

$$v_o(t) = L^{-1}\left[\frac{1/RC}{s+1/RC} \times \frac{V}{s}\right] = V\,L^{-1}\left[\frac{1}{s} - \frac{1}{s+1/RC}\right] = V(1-e^{-t/RC}). \qquad [b6]$$

We may fastly solve even much more intricate cases by using a Laplace transform collection for many functions, and the corresponding inverse transform,.

In conclusion, using Laplace transforms reduces differential equations down to algebra problems, and simplifies the qualitative prediction of the effects of complex transfer functions.

A very short list[111] of the functions most commonly used in electronics, with the corresponding Laplace transform, is given in Figure B.5.

---

[111] An extended list may be found in http://tutorial.math.lamar.edu/Classes/DE/Laplace_Table.aspx; see also http://en.wikipedia.org/wiki/Laplace_transform

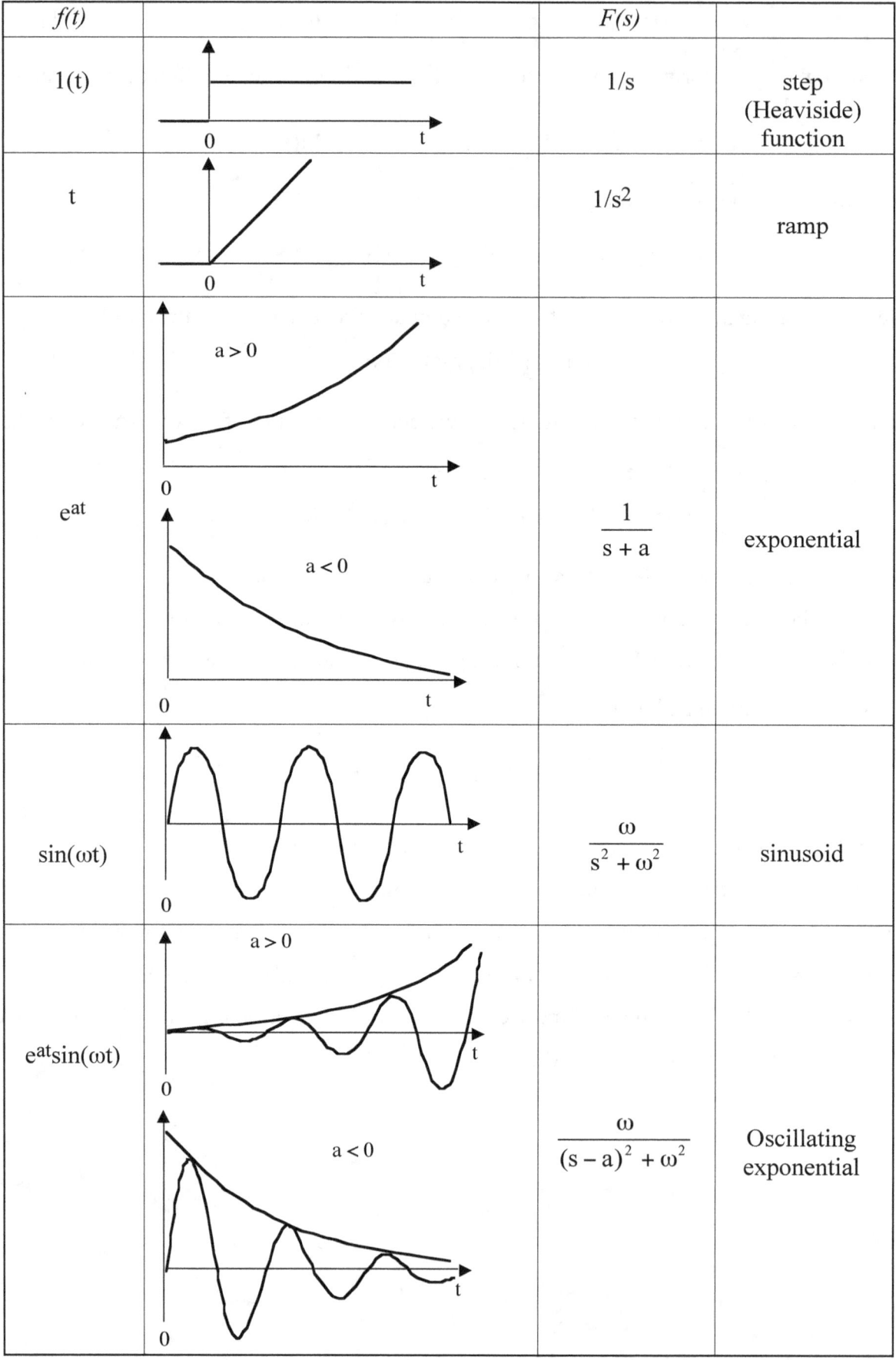

Figure B.5

Appendix C

C.1. Resistors[112]

Resistors are components that dissipate energy: the dissipated power P (by Joule effect) is $P=RI^2=V^2/R=VI$ [watt]. Resistors are commercially available with values in the range from 10 mΩ to 1000 GΩ (i.e. from 10^{-2} Ω a 10^{+12} Ω), and different types may dissipate, without excessive self-heating, power in the range from 1/8 watt to hundreds watt. We may distinguish six main types: carbon-resistors, carbon-film, metal-film, metaloxide-film, wire-wound and foil resistors.

In *carbon-resistors* the resistive element is made from a mixture of finely ground (powdered) carbon and an insulating material (usually ceramic). A resin holds the mixture together. The resistance is determined by the ratio of the fill material (the powdered ceramic) to the carbon. They are available in different sizes that can dissipate power from 0.125 W up to 5 W, and in different types, with tolerances of 3%, 5%, 10% and 20%, with values from 1 Ω to 10 MΩ. They have high temperature coefficient (–0.1%/K) and high electrical noise.

In *carbon film-resistors* the carbon is deposited on an insulating substrate, and a helix cut in it to create a long, narrow resistive path, with usually high value (from 10 Ω to 100 MΩ), good tolerances (0.5%) and lower electrical noise with respect to the normal carbon-resistors.

Wirewound resistors are commonly made by winding a metal wire, usually nichrome, around a ceramic, plastic, or fiberglass core. They are available with values from 1 Ω to 100 kΩ, and tolerances of 1% or better; power dissipated is usually in the range 0.25W-1W (high power models can dissipate up to 200W). General purpose types have high inductance (therefore not suitable for high-frequency applications), but types with anti-inductive winding are also available (at higher cost). The temperature coefficient is normally low (5ppm/K).

Metal film resistors are usually coated with nickel chromium : the resistance value is determined by cutting a helix through the coating rather than by etching. The coating may also be ceramic (cermet) conductors such as (TaN, RuO_2, PbO, $Bi_2Ru_2O_7$, NiCr, or $Bi_2Ir_2O_7$. *Thick film* resistors are manufactured using screen and stencil printing processes. *Thin film* resistors (igher quality, more expensive) are made by sputtering (vacuum deposition) the resistive material onto an insulating substrate; the film is then etched in a similar manner to the old (subtractive) process for making printed circuit boards. They are available in values from 1 Ω to 1000 MΩ, with sizes from 0.25 W to 1 W, with a reasonable tolerance (0.1%, 0.2%, 0.5%, 1%, or 2%) and a temperature coefficient that is generally between 5 and 100 ppm/K; good noise characteristics and low non-linearity due to a low voltage coefficient.

---

[112] See http://en.wikipedia.org/wiki/Resistor

Metal-oxide film resistors. are made of metal oxides such as tin oxide. This results in a higher operating temperature and greater stability/reliability than metal film resistors.

For film or carbon resistors a standard *color code* is used: the first band indicates the first digit of the ohmic value, the second band gives the second digit, the third band gives the exponent in the power of ten multiplier (see Figure C1).

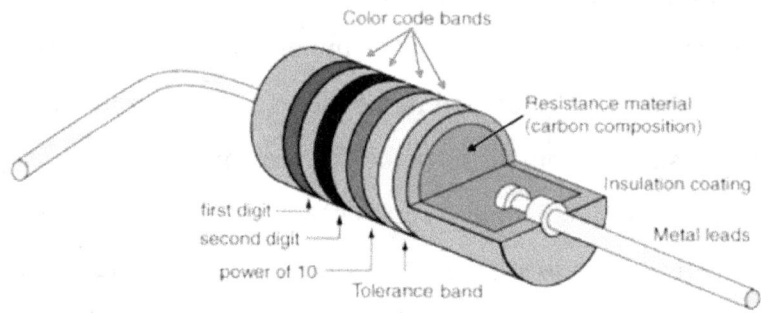

Figure C.1

In Figure C2 the correspondence between colors and digits is shown, as well as the sequence of standard ohmic values commercially available: resistors with 20% tolerance have only the values shown in bold, resistors with 5% tolerance have also values shown in italic.

Color	Digit	Multiplier	Tolerance (%)
Black	0	10^0 (1)	
Brown	1	10^1	1
Red	2	10^2	2
Orange	3	10^3	
Yellow	4	10^4	
Green	5	10^5	0.5
Blue	6	10^6	0.25
Violet	7	10^7	0.1
Grey	8	10^8	
White	9	10^9	
Gold		10^{-1}	5
Silver		10^{-2}	10
(none)			20

sequence of standard values

10 *12* **15** *16* **18** *20* **22** *24* **27** *30* **33** *36*
39 *43* **47** *51* *56* *62* **68** *75* *82* *91* **100**

Figure C.2

Metal-film resistors with tolerance 0.5 % and 1%, use a four digit *numerical code*: the first 3 digits give the value, the fourth gives the mutiplier (power of ten). E.g. 1353 means 135×10^3 Ω. For small values the letter R indicates the decimal (e.g. 10R0 =10.0Ω; 1R0=1.0Ω; R10=0.1Ω).

All resistors have a parasitic capacitance C_p in parallel and some inductance in series L_s (usually negligible below few MHz); the actual impedance of a resistor is therefore $Z=R/(1+j\omega R\, C_p)$.

High value resistors (from 10^9 to 10^{12} Ω) may not be negligible the surface conductivity due to humidity or contaminants (proper degrasing and hydrophobic surface treatment may help)

C.2. Potentiometers

The potentiometer[113] (pot, in electronics slang) is a three terminal component : two end-terminals, and a sweeping contact (wiper) in between. It may be used as voltage divider. By shorting the wiper to one end terminal we get a *rheostat* (a variable resistor, mechanically adjusted) as shown in Figure C3.

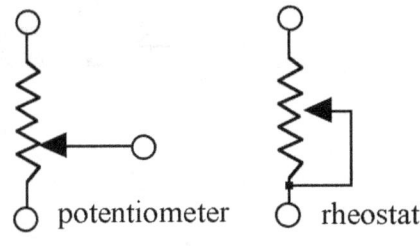

Figure C.3

Most potentiometers have cylindrical geometry with a rotating shaft (6 mm dia) that moves the sweeping contact. In some models the rotating shaft is replaced by a linearly moving wiper, and miniature-size potentiometers (usually panel-mounted or soldered onto printed circuits), named *trimmers*, may be adjusted by a small screwdriver (see Figure C.4).

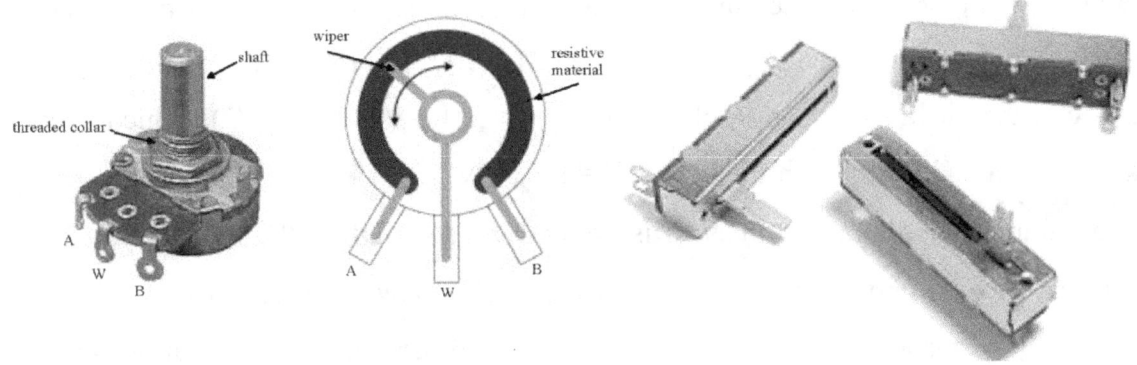

Figure C.4

The resistive path may be carbon-film (cheapest, from 5 Ω to 1 MΩ), conductive plastic or metal wire (most expensive from 10 Ω to 500 kΩ. Wire-wound potentiometers may be multi-turn (*helipot*, with 4, 10, 15, 20 or 25 turns).

C.3. Capacitors

Capacitors[114] are available in a large variety of shapes and types. The specifications for a capacitor usually include the value of capacitance C, the voltage rating (i.e. the maximum voltage which can be continuously applied), the temperature coefficient, the leakage current I_p (or leakage resistance R_p), the dissipation factor DF.

The capacitance is expressed by the relation $C=K\varepsilon_0 A/d$, where *A* and *d* are the electrode's area and separation, ε_0 the permittivity of free space and *K* the relative permittivity (or dielectric constant) of the material separating the electrodes. Therefore large C values imply large *A* (i.e.

---

[113] See http://en.wikipedia.org/wiki/Potentiometer
[114] See http://freecircuits.org/2012/01/capacitors-basics-working/

large dimensions) and small d (i.e. maximum voltage limited by dielectric breakdown).

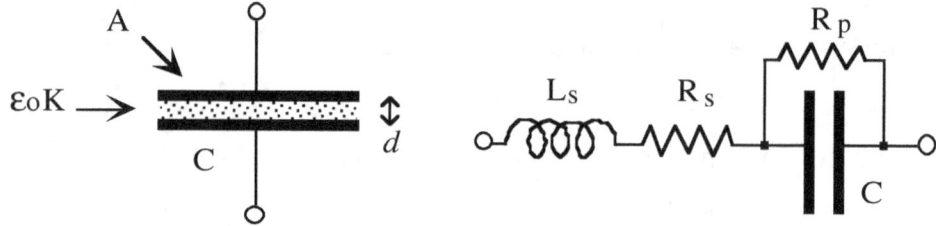

Figure C.5

The *leakage resistance* R_p is the ratio between the applied voltage and the *leakage current* I_p: it is the effective resistance in parallel to the capacitor, and it is normally proportional to $1/C$ (for a given K the product R_pC is constant, and measured in seconds or $M\Omega \times \mu F$).

The complex impedance of a real capacitor may therefore be written as $Z_c = R_p/(1+j\omega\, R_pC)$.

A capacitor has also a *series resistance* R_s, and a *series inductance* L_s, usually negligible ($R_s < 1\ \Omega$, but sometimes much larger).

The *dissipation factor*[115] DF, measured in sinusoidal regime, is the ratio between the energy dissipated and the energy stored within one cycle (DF = 1/Q, where Q is the quality factor) and it is nearly constant in a wide range of frequency f : DF = (P /f)/(CV²/2), which means that dissipation increases linearly with frequency: P = DF (C V²/2) f.

A related parameter is the *loss tangent* δ is defined as $\tan\delta = DF$ is the ratio between real and imaginary parts tof the capacitor impedance. Ideal capacitor have $\delta = 0$ ($R_p = \infty$).

Capacitors may be made made of two conducting films, separated by an insulating film, spirally wound into a compact cylinder, or by a ceramic disc with two metal plates on opposite surfaces. The symbols used for capacitors distinguish normal, elecrolutic and tunable (Figure C.6)

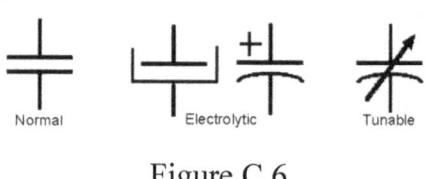

Figure C.6

Different models are distinguished by the type of insulating spacer: air, ceramic, mica, polystyrene, polyester (PET, mylar), polymide (Kapton), polycarbonate (KC), polypropylene, PTFE (Teflon), aluminum oxide, tantalum oxide, oil, paper, glass[116].

Air-gap capacitors have a low dielectric loss. Used mainly for large-valued, *tunable* capacitors that can be used for resonating HF antennas .

Ceramic and mica capacitors (due to the low K value) have small capacitance but also small DF; they are useful in high frequency circuitry. They are marked by a color-code similar to that for resistors, with some differences (Figure C.7).

---

[115] See http://en.wikipedia.org/wiki/Dissipation_factor
[116] See http://en.wikipedia.org/wiki/Types_of_capacitor

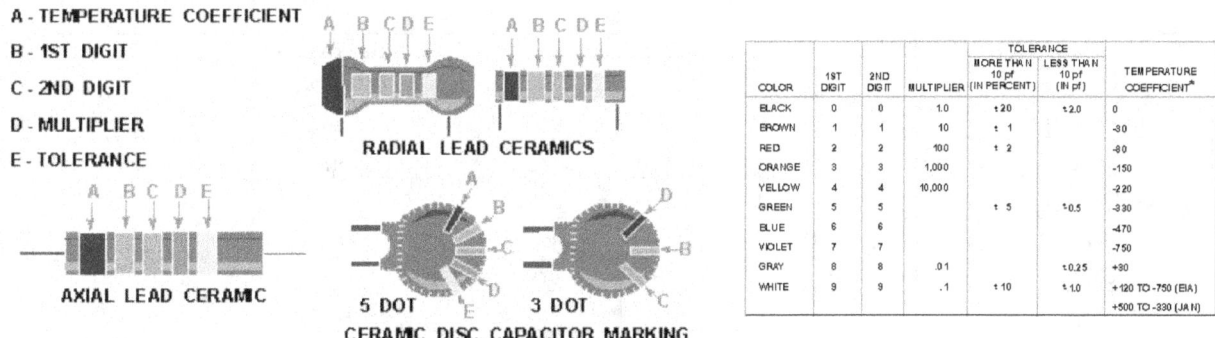

Figure C.7

Among plastic-film capacitors the *polyester*-type (*mylar*: K=3.1, DF≈0.3) have high voltage rating, the *polystyrene*-type (K=2.5) have small voltage rating but very small dissipation (DF≈0.03); the *polycarbonate*-type (K=2.8) have small temperature dependence (TC≈+300 ppm./K) more details are shown in Table C.1.

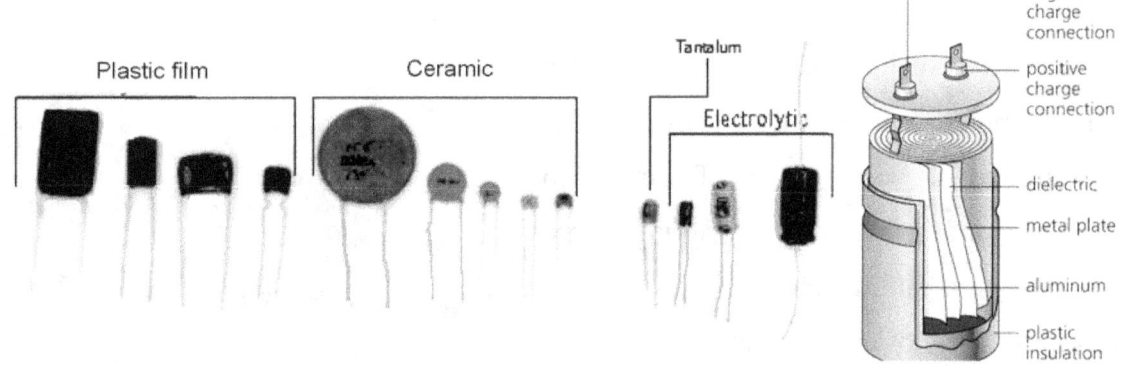

Figure C.8

Electrolytic capacitors (Figure C.8) require the application of a DC bias voltage in order to work properly. This voltage must be applied with the correct polarity (invariably this is clearly marked on the case of the capacitor) with a positive (+) sign or negative (−) sign or a coloured stripe or other marking. Failure to observe the correct polarity can result in over-heating, leakage, and even a risk of explosion. They have high capacitance/volume ratio, but high leakage. There are also non-polar types made by two capacitors in series with reversed polarity. May be made of aluminum-oxide or tantalum-oxide (more reliable but more expensive). The value of capacitance is written in µF units, even if sometimes the letter µ is replaced by "m", and also the maximum voltage is marked. Also some paper capacitors (not polar) have a mark indicating the external foil that should be grounded to minimize the pick-up noise.

Some properties of variour capacitors are listed in Table C.1 [the units are *Farad* (F), with sub-units : p= pico = 10^{-12}, n= nano = 10^{-9}, µ= micro = 10^{-6}]

Type	Values	V max (V)	Temp.coeff.	Leakage resistance	Comments
mica/glass	1pF-10nF	100-600	very good	good	hf, expensive
ceramic	10pF-100nF	50-3000	fair	fair	hf, cheap, small
Polymide (Kapton)	1nF-10µF	10 kV	fair	good	up to 250°C
polystyrene	10pF-1µF	100-300	good (negative)	excellent	for filters
polycarbonate	100pF-30µF	50-800	very good	good	large size,
polypropylene	100pF-50µF	50-300	good	excellent	low DF
polyester (Mylar)	1nF-2µF	10 kV	fair	good	up to 125°C
teflon (PTFE)	1nF-2µF	200	good	best	up to 250°C
tantalum	100nF-1000µF	6-100	fair		small size
aluminum	100nF-0.001F	3-600	bad	bad	high C, cheap
oil	0.1 µF-20µF	–>10.000	faif	good	large size

Table C.1

C.4. Inductors

An inductor[117] is usually constructed as a coil of conducting material, typically copper wire, wrapped around a core either of air or of ferromagnetic or ferrimagnetic material. Cores with a higher permeability increase the magnetic field and confine it closely to the inductor, thereby increasing the inductance. Low frequency inductors are constructed like transformers, with

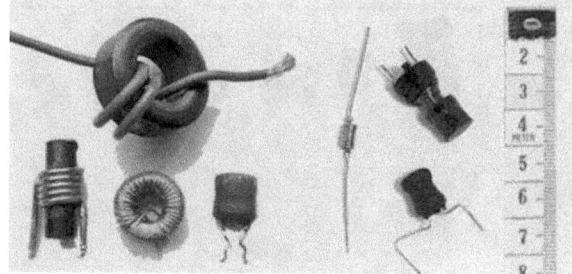

Figure C.9

cores of electrical steel laminated to prevent eddy currents. *Ferrites* (ceramics filled by iron oxide) are widely used for cores above audio frequencies, since they do not cause the large energy losses at high frequencies that ordinary iron alloys do. Inductors come in many shapes. Some inductors have an adjustable core, which enables changing of the inductance.

Inductors have always a parasitic resistance R_p in series and a parasitic capacitance in parallel (usually negligible), so that the inductor impedance may be written: $Z_L \approx R_p + j\omega L$.

It is measured in Henry: commercially available values for air-core inductors are in the range from 0.01µH to some millihenry (mH), while ferrite-core inductors may have values from 1 µH up to several henry (with $R_p \approx 100\Omega$).

---

[117] See http://en.wikipedia.org/wiki/Inductor

C.5. Diodes

Usually diodes are marked by a line on the cathode side (N). There are several different types: *signal diodes* (low power <1W) with small reverse current (of the order of µA, some below 1 nA), *rectifying diodes* for large forward currents (up to 100 A) with larger reverse currents (some mA). The fast rectifiers (*switching diodes*) have short recovery time (for emptying the junction depletion layer): 1N4148, 1N4150, 1N4151, 1N4448, 1N914,1N916 have reverse current smaller than 0.1µA;. *Schottky diodes* (e.g. BAR10, BATxx, HSCH1001, 1N5712, 1SS108) are constructed from a metal to semiconductor contact. They have a lower forward voltage drop than p–n junction diodes, in the range 0.15 V to 0.45 V, and they have a faster reverse recovery than p–n junction diodes, they are recommended for small signals, high frequency.

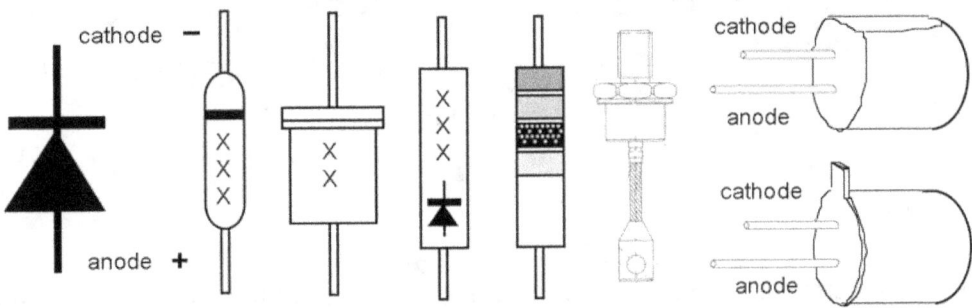

Figure C.10

Table C.2 lists some common diode characteristics: V_B = maximum reverse voltage (*breakdown* voltage), I_o = reverse current (or *leakage* current), V_F = forward voltage drop I_d = forward current, C = parasitic capacitance.

Name	V_B (V)	I_o (µA)	V_F (V)	I_d (mA)	C (pF)	Comments
FJT1100	30	.001	1.1	.05	1.2	low I_O
1N3595	150	3	0.7	10	8	(fast) low I_O
1N914	75	5	.75	10	4	signal (fast)
1N4148	75	5	.75	10	4	signal (fast)
1N456/9	30/200	0.025	1	40/3		(fast) low I_O
1N6263	60	10	.4	1	1	(fast) low V_f
1N3062	75	50	1	20	.6	(fast) low C
1N4002	100	50	.9	1000	15	rectifier 1A
1N4007	1000	50	.9	1000	10	rectifier 1A
1N5625	400	50	1.1	5000	45	rectifier 5A
1N1183A	50	1000	1.1	40000		rectifier 100A peak

Table C.2

Zener diodes may have breakdown voltages V_z in the range from 250 mV to 1.5 kV.

The series Semtech BZV85CxxVx gives many V_z values (V_z=2.7, 4.7, 5.1, 5.6, 6.8, 7.5, 8.2, 8.1,

10, 11, 12 ...) up to 200 V. Their name xxVx stays for the V_z value in volt, where V is the decimal point. The minimum reverse current I_z increases with decreasing V_z. For small I_z (0.05 mA) the Texas series 1N4678÷1N4700 goes from $V_z = 1.8$ V up to 25 V: the value increases with the number: 1N4679 = 2.0 V, 1N4680 = 2.2 V etc.

A better temperature stability is achieved by bandgap zeners as the series LM103XX, or LM199, LM329, LM113, AD589... (see also § 13.2).

In the Light Emitting Diodes (LED) the cathode is marked by the flat side of the cap, or by a stripe in metal can, or by the shorter lead (Figure C.10).

Rectifying diodes may be available pre-assembled into *Graetz bridge rectifiers* (Figure C.11)

Figure C.11

C.6. Solderless breadboard

In order to test an electronic circuit without using soldered junctions we may use a *solderless breadboard*, that does not require soldering, and is reusable. Moreover it makes easier changing the circuits or replacing components without risks of overheating.

A modern solderless breadboard consists of a perforated block of plastic with numerous tin plated phosphor bronze or nickel silver alloy spring clips under the perforations. The clips are often called *tie points* or *contact points*.

The spacing between the clips (*lead pitch*) is typically 2.54 mm. Integrated circuits (ICs) in dual in-line packages (DIPs) can be inserted to straddle the centerline of the block. Interconnecting wires and the leads of discrete components can be inserted into the remaining free holes to complete the circuit. Typically the spring clips are rated for 1 ampere at 5 volts and 0.3 amperes at 15 volts (5 watts).

The layout of a typical solderless breadboard is made up from two types of areas, called *Terminal strips* and *Bus strips*. *Terminal strips* are the main areas, to hold most of the electronic components. In the middle of a terminal strip of a breadboard, one finds a *notch* running in parallel to the long side that marks the centerline and provides airflow (cooling) to DIP ICs straddling the centerline. The clips on the right and left of the notch are each connected in a radial way; typically five clips in a row on each side of the notch are electrically connected.

Bus strips provide power to the electronic components. A bus strip usually contains two columns: one for ground and one for a supply voltage. Bus strips typically run down both sides of a terminal strip.

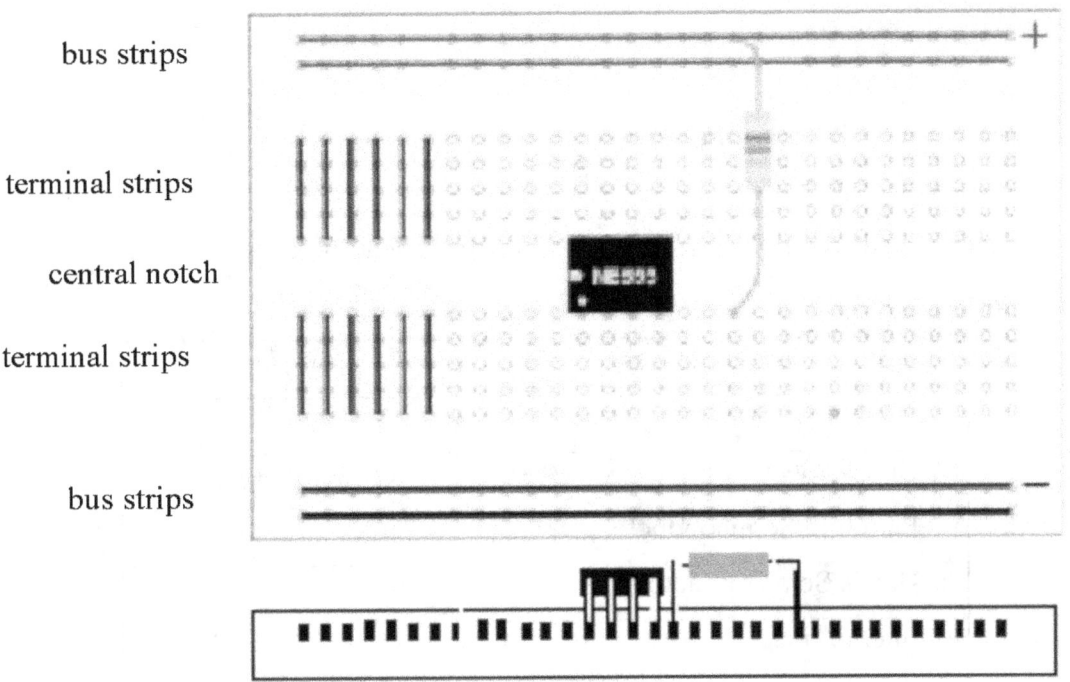

Figure C.12

Appendix D

Commercial IC: characteristics and pin-out

D.1. Short list of linear IC manufacturers

The first characters of the name in a device give information about the manufacturer.

Manufacturer	Initials
Analog Devices Inc.	AD
Burr-Brown	OPA-(none)
Fairchild Semiconductors	µA
Harris Semiconductors	HA-(CA)
Intersil Inc.	ICL-ICM-FLT
Linear Technology	LT
Maxim	MAX-(BB-ICL)
Motorola Semiconductors	MC-(LF-LM-TL)
National Semiconductors Corp.	LF-LH-LM
Precision Monolitics Inc.	OP
Raytheon Semiconductors	RC-RM
RCA Solid State Division	CA-CD
Sprague	ULN-ULS-ULX
Siliconix	L
Signetics Corp.	NE-SE -SU
SGS-Ates	LS
Texas Instruments Inc.	SN-TL-TLC-(µA)

D.2. Pin-out and general data sheets of Operational Amplifiers

The pinout identification is found in each datasheet, but some general features are the following. Pin numbers run always in clockwise direction (top-view), and pin 1 is the closest to the marker. The marker (see figure) is a dot, or notch in plastic model and a tab in metal can.

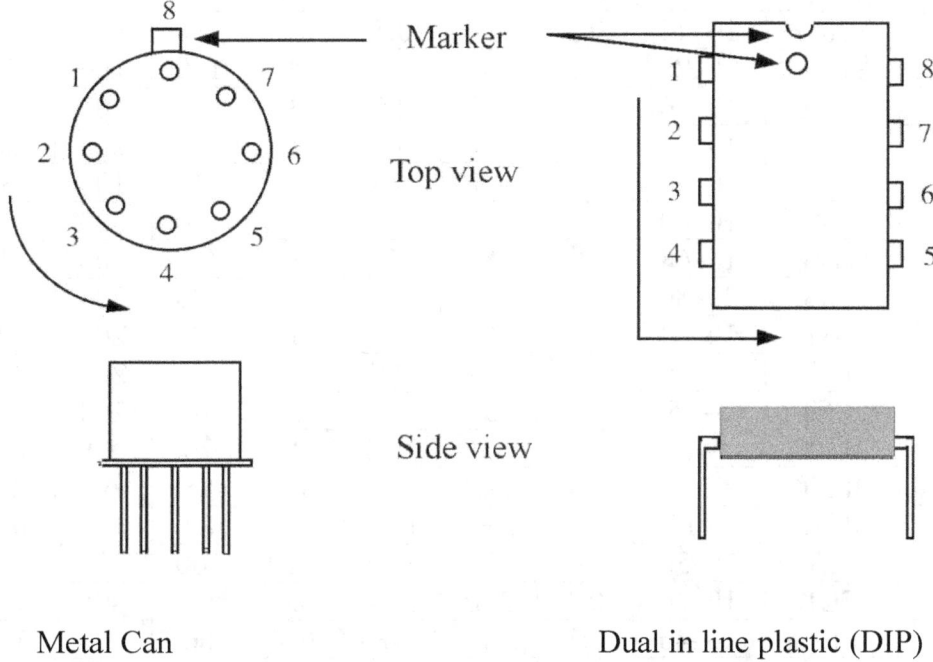

Metal Can   Dual in line plastic (DIP)

The characteristics of OA are given by manufacturers within *Data Sheets*, freely downloadable in internet, that often include useful suggestions for circuit design[118].

In the following tables we list some data for the most common IC: the commercially abvailable devices are thousands and new models are continuously produced.

---

[118] Texas: http://www.ti.com/ww/en/home/three-col/
Analog Devices: http://www.analog.com/en/amplifiers-linear/products/index.html,
Fairchild: http://www.fairchildsemi.com/products/
BurrBrown: http://www.burrbrown.info/ and http://www.datasheetcatalog.net/it/burrbrown/1/
Maxim:.http://www.maxim-ic.com/design/techdocs/app-notes/

D.2.1. OA with pinout "741"

Name	input stage	V_{cc} (V) Min-Max	V_{os} (mV)	I_b (nA)	I_{os} (nA)	ω_t (MHz)	CMMR (dB)	A (10^3)	I_s (mA)	I_o (mA)
µA741	bipolar	10–36	2	80	20	1.2	90	200	2.8	20
AD741	bipolar	10–44	0.5	30	2	1	110	200	2.8	15
LS148	bipolar	4–22	1	80	20	1	90	150	1.9	25
OP01	bipolar	10–44	1	20	1	2.5	100	100	3	6
OP02	bipolar	10–44	0.3	18	0.5	1.3	100	250	2	6
RC4131	bipolar	7–36	1.5	70	3	4	100	160	2	10
NE530	bipolar	10–36	2	65	15	3	90	200	3	10
NE535	bipolar	10–36	2	65	15	1	90	200	2.8	10
MC1456	bipolar	10–36	5	15	5	1	110	100	3	5
MC1436	bipolar	10–80	5	15	5	1	110	500	5	10
LM143/343	bipolar	10–68	2	8	1	1	90	100	2	20
HA2645	bipolar	20–80	2	15	12	4	100	200	4.5	10
MC1741	bipolar	10–44	6	200	30	1	90	200	3.5	10
TL081	JFET	5–15	5	.03	.005	3	80	200	2.8	10
TL071	JFET	5–15	3	.03	.005	3	76	200	2.5	10
TL061	JFET	5–15	3	.03	.005	1	76	10	0.25	5
TL051	JFET	5–15	0.7	20pA	4 pA	3.6	85	60	2.3	80
TL031	JFET	5–15	0.5	2 pA	1 pA	1	87	7	0.2	40
LF351	FET	10–36	5	.05	.025	4	100	100	3.4	10
AD515	FET	10–36	0.4	.3 pA	.3 pA	0.4	94	40	1.5	10
3528BM	FET	10–40	0.1	.2 pA	.04pA	0.7	86	100	1.5	10
CA3140	Mosfet	4–44	2	10pA	.5 pA	3.7	90	100	6	10
CA3160	Mosfet	5–16	2	5 pA	.5 pA	4	90	320	15	12
OPA602	DiFET	5-18	.25	1 pA	.5pA	6.5	100	100	20	4

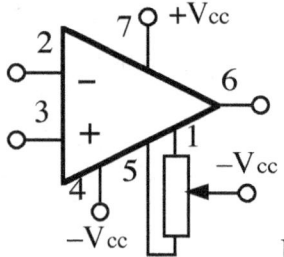

Pins 1-5 are for offset null

D.2.2. OA with pinout "356"

Name	input stage	V_{cc} (V) Min-Max	V_{os} (mV)	I_b (nA)	I_{os} (nA)	ω_1 (MHz)	CMMR (dB)	A (10^3)	I_s (mA)	I_o (mA)
LF355	FET	10–36	3	0.03	0.003	2.5	100	100	4	20
LF356	FET	10–36	3	0.07	0.007	4.5	100	100	10	20
LF357	FET	10–36	3	0.07	0.007	20	100	100	10	20
OP15	FET	10–44	0.2	0.015	0.003	6	100	240	4	15
OP16	FET	10–44	0.2	0.015	0.003	8	100	240	7	20
AD825	FET	5–15	1	0.02	0.02	41	80	6.5	6	50
LM110 /210/310	bipolar	5–18	1.5	1.5	10	20	100	1	4	5
LM112 /212/312	bipolar	5–18	1	1	1	0.3	100	20	0.3	5
LM216 /316	bipolar	5–20	0.5	5	0.05	0.1	80	30	0.6	5
AD504	bipolar	10–36	0.5	80	2	0.3	110	1000	3	15
AD510	bipolar	10–36	0.02	10	-	0.3	110	1000	3	10
AD517	bipolar	10–36	0.02	5	3	0.25	100	1000	4	10
µA725	bipolar	5–20	0.5	42	2	0.08	100	3000	3	5
OP05	bipolar	6–44	0.2	1.2	1.2	0.6	123	500	4	10
OP07	bipolar	6–44	0.01	0.7	0.3	0.6	126	500	4	10
HA2500 /02/05	bipolar	10–20	4	100	20	0.5	90	60	4	10
HA2510 /12/15	bipolar	10–20	4	100	20	0.5	90	15	4	10
HA2520 /22/25	bipolar	10–20	4	100	20	2	90	15	4	10
OPA177	bipolar	10–15	0.01	1	1	0.6	60	6000	1	20

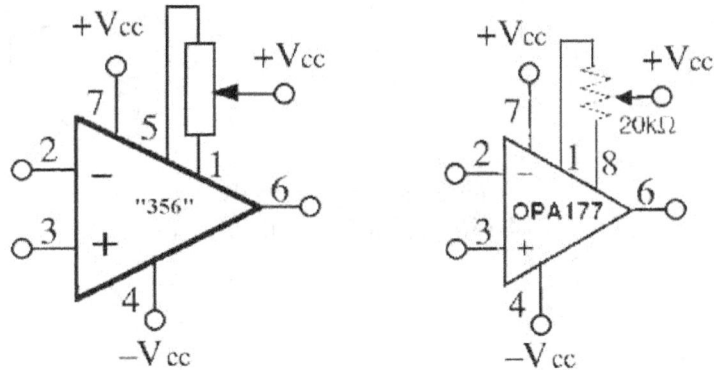

offset null pins: either 1-5 or 1-8

D.2.3. Dual operational amplifiers

Name	input stage	V_{cc} (V) Min-Max	V_{os} (mV)	I_b (nA)	I_{os} (nA)	ω_1 (MHz)	CMMR (dB)	A (10^3)	I_s (mA)	I_o (mA)
MC1458	bipolar	5–18	2	80	20	1.2	90	200	2.8	20
RC4558	bipolar	5–15	1	40	5	3	100	300	7	20
LM158/258/358	bipolar	3–18	2	50	10	1	90	200	3	20
µA798 (#)	bipolar	3–18	2	50	10	1	90	200	3	20
OP04 (*)	bipolar	5–22	0.3	18	0.5	1.3	100	250	2	6
OP14	bipolar	5–22	0.3	18	0.5	1.3	100	250	2	6
OPA2604	FET	4–24	1	0.1	0.004	20	100	100	35	12
µA747 (**)	bipolar	5–18	2	80	20	1.2	90	200	2.8	20
TL082	JFET	5–15	5	.03	.005	3	80	200	2.8	10
TL072	JFET	5–15	3	.03	.005	3	76	200	2.5	10
TL062	JFET	5–15	3	.03	.005	1	76	10	0.25	5
TL052	JFET	5–15	0.7	20pA	4 pA	3.6	85	60	2.3	80
TL032	JFET	5–15	0.5	2 pA	1 pA	1	87	7	0.2	40
LF353	JFET	5–18	5	.05	.025	4	100	100	3.4	10
µA772	JFET	5–18	2	.05		3	80	100	3	10

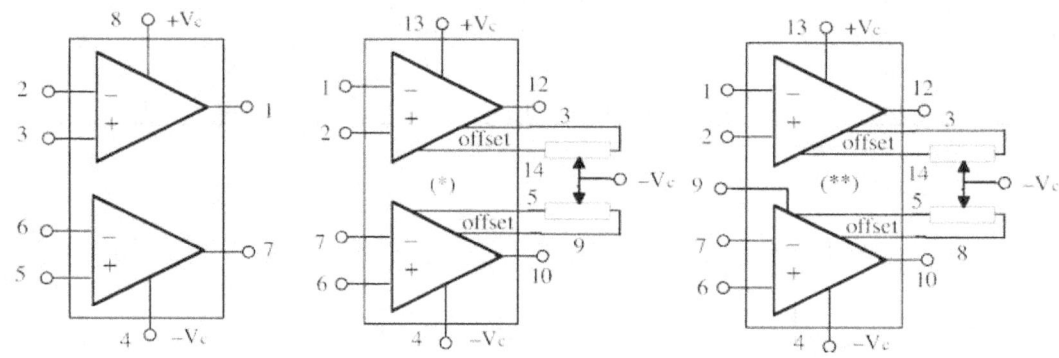

(#) accepts single supply
(*)(**) 14-pin available, with offset null

D.2.4. Quad operational amplifiers

Name	input stage	V_{cc} (V) Min-Max	V_{os} (mV)	I_b (nA)	I_{os} (nA)	ω_t (MHz)	CMMR (dB)	A (10^3)	I_s (mA)	I_o (mA)
MC4741	bipolar	5–18	2	80	20	1.2	90	200	2.8	20
RC4156	bipolar	3–20	5	60	30	3.5	80	100	7	20
LM148/248/348	bipolar	5–18	2	80	20	1.2	90	200	2.8	20
LM124/224/324	# bipolar	3–30	2	45	5	1	100	50	0.8	30
OP11	bipolar	5–22	0.5	300	25	2	120	600	3	15
TL084	JFET	5–15	5	.03	.005	3	80	200	2.8	10
TL074	JFET	5–15	3	.03	.005	3	76	200	2.5	10
TL064	JFET	5–15	3	.03	.005	1	76	10	0.25	5
TL054	JFET	5–15	0.7	0.02	0.004	3.6	85	60	2.3	80
TL034	JFET	5–15	0.5	0.002	0.001	1	87	7	0.2	40
LF347	JFET	5–18	3	0.05	0.025	3	100		7	
µA774	JFET	5–18	10	0.2	0.1	3	70	25	3	25

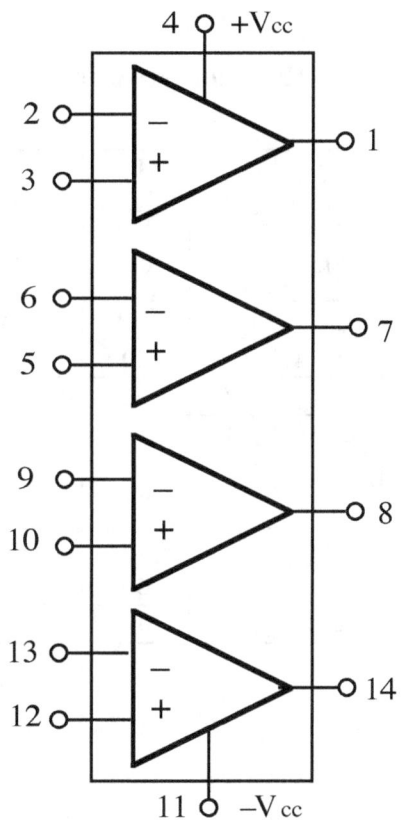

(#) accepts single supply

D.2.5 Instrumentation amplifiers

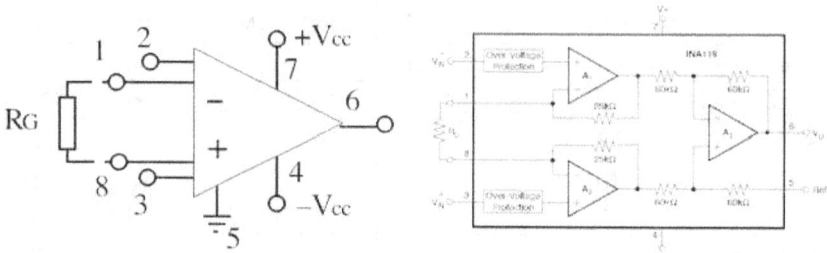

INA114, INA115, INA118, OPA2604, OPA177 :
the gain is adjusted by the external resistor R_G: $G = 1 + 50k\Omega / R_G$

D.3. Comparators

Name	open collector	V_{cc} (V)	V_{os} (mV)	I_b (μA)	I_{os} (μA)	τ_s (μs)	Num Comp.	Single Supply
µA111/311 LM111/211	yes	±15	1	0.1	0.04	0.2	1	yes
LF111/211/311	yes	±15	4	50 nA	.02 nA	0.2	1	yes
µA710-LM710	no	−7+14	0.6	20	3	0.04	1	no
LM106/206/306	yes	±12	2	20	3	0.04	1	no
LM119/219/319	yes	±15	4	0.5	0.1	0.08	2	yes
LM139/239/339 µA139/239/339	yes	±18	2	0.2	0.05	1.3	4	yes
LM193/293/393	yes	±18	1	0.1	0.02	1.3	2	yes
µA711-LM711	no	−7+14	3	50	10	0.04	2	no
LP165/365	yes	±18	3	0.1	0.05	4	4	yes

D.4. Basic list of logic gates (TTL and CMOS)[119]

Type	number of inputs	number of gates	TTL drawing	TTL name	CMOS drawing	CMOS name
Inverter	1	6	a	7404	a	4069
AND	2	4	e	7408	c	4081
AND	3	3	g	7411	h	4073
AND	4	2	e	7421	f	4082
NAND	2	4	b	7400	c	4011
NAND	3	3	g	7410	h	4023
NAND	4	2	e	7420	e	4012
NAND	8	1		7430		4068
OR	2	4	b	7432	c	4071
OR	3	3		–	h	4075
OR	4	2		–	f	4072
NOR	2	4	d	7402	c	4001
NOR	3	3	g	7427	h	4025
NOR	4	2		7425	e	4002
XOR	2	4	b	7486	c	4070
XNOR	2	4		74266	c	4077
Schmitt NAND	2	4		–	c	4093
Schmitt Inverter	1	6	a	7414	a	4584

[119] Pinouts of 74xx series may be found in http://www.romux.com/pinouts/74-series/pin-identification

Bibliography

The number of textbooks devoted to electronics teaching is quite large. However it is not easy to find textbooks on this subject which are at the same time rigorous and *"reader-friendly"*: in the following bibliography we tried to collect a short list of "good textbooks" that should be sufficient to satisfy the curiosity left by the present book to the most demanding readers.

1) *The Art of Electronics*, P. Horowitz and W. Hill, Cambridge Univ. Press, 1980, 700 pages. A fundamental textbook for learning analog and digital electronics. Offers practical examples, rules of thumb, and a large bag of tricks a largely nonmathematical treatment that encourages circuit intuition, brainstorming, and simplified calculations of circuit values and performance. *The Art of Electronics-Student Manual*, T. Hayes and P. Horowitz : provides extra explanatory notes, worked examples, solutions to selected exercises and laboratory exercises.
2) *Microelectronics*, J. Millman and A. Grabel, Mc Graw Hill, 1987, 1000 pages. Suggested for more detailed discussion of transistors and digital devices. Interesting chapters devoted to discrete elements amplifiers, to A/D D/A conversion, and to power amplifiers.
3) *Operational Amplifiers*, G. B. Clayton, S. Winder , Newnes 2003, 397 pages. One of the most successful books written on Operational Amplifiers now revised and fully updated (fifth ed.).
4) *Linear Integrated Circuit Applications*, G. B. Clayton, Macmillan, 1975, 270 pages. An extension of previous book with chapters on active filters, on timers and on transconductance multipliers/dividers.
5) *Experiments with Operational Amplifiers*, G. B. Clayton, Macmillan, 1975, 120 pages. Good, short collection of examples and exercises on OA.
6) *Linear Circuit Design Handbook*, H. Zumbhalen, Newnes, 2008, 943 pages. A very good textbook including chapters on sensor interfacing.
7) *Operational Amplifiers: Design and applications*, J. G. Graeme, G. E. Tobey and L. P. Huelsman, McGraw Hill, 1971, 470 pages. Book of the Burr-Brown series; detailed characteristics of real OA, rich collection of circuit examples, active filters discussion.
8) *Introduction to Operational Amplifiers: Theory and Applications,* J. Wait, L. Huelsman and G. Korn, McGraw Hill, 1975, 390 pages. Very nice chapter on non-linear functions and multipliers
9) *Applications of Operational Amplifiers*, J. G. Graeme, McGraw Hill, 1973, 230 pages. Book of the Burr-Brown series; includes a very rich collection of practical circuit examples
10) *Lock-in Amplifiers: Principles and Applications*, M. L. Meade, Peregrinus Ltd., 1983, 230 pages. Unique monograph on lock-in amplifiers: clear and exhaustive, guides the from first models to modern .
11) *TTL Cookbook* , D. Lancaster, Howard W. Sams, 1981, 330 pages. Illustrates how TTL is used in many practical applications; provides typical circuits and working applications, including a rich collection of TTL pinouts.
12) *CMOS Cookbook*, D. Lancaster, Howard W. Sams, 1981, 410 pages. Requires little math, this practical, user-oriented book covers all the basics for working with digital logic.
13) *Digital Electronics, an hands-on learning approach*, W. G. Young, Hayden Book Co., 1980, 200 pages. A textbook of digital electronics featuring almost exclusively an experimental or laboratory approach.
14) *IC Timer Cookbook* , W. G. Young, Howard W. Sams, 1981, 200 pages. The greatest book written on the NE555. Thirty years later this book is still outstanding.
15) *Elementary Semiconductor Physics*, H. C. Wrigth, Van Nostrand, 1979, 77 pages. A booklet that offer a compact introduction to the physics involved in most semiconductor devices (p-n junction, metal-semiconductor junction,...)
16) *The Physics of Semiconductor Devices*, D. A. Fraser, Clarendon Press, 1983, 170 pages. The book emphasizes diagrams rather than complicated analytical methods

17) Introduction to Semiconductor Physics, R. Adler, A. Smith, and R. Longini, John Wiley & Sons, 1964, 250 pages. An introductory handbook that includes some guided experiments in the semiconductor physics.
18) Op Amp Applications Handbook, Walt Jung, Newnes, 2006, 283 pages; AnalogDevices series; Comprises Op Amp Basics, Specialty Amplifiers, Using Op Amps with Data Converters, Sensor Signal Conditioning, Analog Filters, Signal Amplifiers, Hardware and Housekeeping Techniques, and Op Amp History.
19) *Semiconductor Controlled Rectifiers*, F. Gentry, F. Gutzwiller, N. Holonyak and E. von Zastrov Prentice Hall, 1965, 380 pp. Book devoted to principles and applications of p-n-p-n device (SCR, TRIAC).
20) *Transducers in Mechanical and Electronic Design*, H. Trietley, Dekker Inc., 1986, 370 pp. Provides information on the features of various transducers, (temperature, pressure, position, flow, vibration, shock, acceleration, conductivity, pH,...).
21) *Electronics for the Physicist*, C. G. Delaney, Ellis Horwood Ltd., 1980, 300 pp. Aimed at teaching particles detectors electronics, may be useful to a broader range of readers. Nice chapters on Laplace transform and on electrical noise.
22) *Instrument Transducers*, H. K. Neubert, Oxford University Press, 1963, 390 pages. An old book, still valid: with a smart treatment of semiconductor physics related to transducers
23) *Electronic Circuits and Applications*, S. Senturia and B. Wedlock, J. Wiley, 1975, 600 pages.
24) *Feedback and control system analysis and synthesis*, J. J.D'Azzo and C.H. Houpis, Mc Graw Hill, 1986, 824 pages. Interesting for the analysis of stability criteria.
25) *Handbook Of Operational Amplifier Active RC Networks*, B. Carter and L.P. Huelsman Texas Instrument Application Report, SBOA093A ; 85 pages, 2001.
26) *Handbook Of Operational Amplifier Applications* Bruce Carter and Thomas R. Brown, Texas Instrument Application Report, SBOA092A, 94 pages, 2001
27) *A Current Feedback Op-Amp Circuit Collection*, B. Carter, Texas Instrument Application Report SLOA066; 10 pages, 2001
28) *Op Amps For Everyone*; 1st Ed; Ron Mancini; 464 pages; 2002; Texas Instruments Application Report SLOD006B.
29) *Physics of Semiconductor Device*, S. M. Sze and Kwok Kwok Ng, John Wiley, 2007, 815 pages. Standard textbook and reference in the field of semiconductor devices.
30) *Semiconductor Devices: Physics And Technology*, 2Nd Ed, S.M.Sze; Wiley India, 2008 - 572 pages; Provides strong coverage of all key semiconductor devices. Includes basic physics and material properties of key semiconductors· Covers all important processing technologies
31) *Op-Amps and Linear Integrated Circuits*; 4th Ed; Ram Gayakwad; 543 pages; 1999; An easy-to-understand book that presents the basic principles of OA and IC, with a practical approach.
32) *Operational Amplifiers and Linear Integrated Circuits*; 6th Ed; Robert F Coughlin; 529 pages; 2000 Presents a clear approach for op-amp courses TTL, and more.
33) *Basic Operational Amplifiers and Linear Integrated Circuits*; 2nd Ed; Thomas L Floyd; David Buchla; 593 pages; 1998; The book integrates theory, practical circuits, and troubleshooting concepts, keeping mathematical details to a minimum.
34) *Design with Operational Amplifiers and Analog Integrated Circuits*; 3rd Ed; Sergio Franco; 672 pages; 2002 This text emphasizes the physical approach; there is a "building block approach" to deal with the harder concepts and 176 worked examples.. PSPICE is used as an analysis tool.
35) *Active Filter Cookbook*, D. Lancaster, Howard W. Sams & CO, 240 pages ;1975, ISBN: 0-672-21168-8 ; includes practical elements such as working circuits, ready-to-use design tables, tuning, and real-world applications, making it easy to use and apply
36) *Introductory Experiments in Digital Electronics*, by H.M.Berlin (Book 1&2), Howard W. Sams & Co., Inc. Publisher SAMS
37) *Logic & Memory Experiments using TTL Integrated Circuits*, by H.M.Berlin (Book 1&2), Howard W. Sams & Co., Inc. Publisher SAMS.

38) *The Design of Active Filters, with Experiments*, by H.M.Berlin, 1974. Howard W. Sams & Co., Inc. Publisher SAMS.
39) *Understanding IC Operational Amplifiers,* by R. Melen and H. Garland, 1971. Reston Publishing Co., Publisher RESTON
40) *Sensor Technology Handbook* Jon S. Wilson , Newnes, 700 pages 2005 ;In addition to background information on sensor technology, measurement, and data acquisition, the handbook provides detailed information on each type of sensor technology, covering: technology fundamentals sensor types. The handbook also contains information on the latest MEMS and nanotechnology sensor applications
41) *Sensors and Transducers,* Ian Robertson Sinclair, Newnes, May 30, 2001 - 306 pages The focus is on understanding the technologies and their different applications, not a mathematical approach. A text which provides an introduction to the selection and application of sensors, transducers and switches, and a grounding in the practicalities of designing with these devices. The devices covered encompass heat, light and motion, environmental sensing, sensing in industrial control.

The author

Giacomo Torzo, physicist now retired, was research director at *Consiglio Nazionale delle Ricerche* (CNR) since 1993, while teaching Physics Laboratory at Padova University until 2011.

www.ingramcontent.com/pod-product-compliance
Lightning Source LLC
Chambersburg PA
CBHW081046170526
45158CB00006B/1872